Agriculture Development and Capital Relation: A Modern View

Agriculture Development and Capital Relation: A Modern View

Mahesh Khurana

CYBER TECH PUBLICATIONS
4264/3, Ansari Road, Daryaganj, New Delhi-110002 (India)
Ph.: 011-23244078, 011-43559448 Fax: 011-23280028
E-Mail: cyberpublicationsdelhi@yahoo.com
Website: www.cybertechpublications.com

AGRICULTURE DEVELOPMENT AND CAPITAL RELATION: A MODERN VIEW

Mahesh Khurana

First Edition 2010

Published by :

G.S. Rawat for **Cyber Tech Publications**
4264/3, Ansari Road, Daryaganj, New Delhi-110002 (India)
Ph.: 011-23244078, 011-43559448 Fax: 011-23280028
E-Mail: cyberpublicationsdelhi@yahoo.com
Website: www.cybertechpublications.com

Preface

The crisis of the 1970s and 1980s results from the combined effects of the over-exploitation of the land and of men and women, raised to a level that can be raised no more, and the crisis engulfing the capitalist system as a whole. Faced with this the proposals that rain down on Africa at an ever increasing rate are scarcely anything more than the expression of the 'quest for palliatives', as Gakou shows.

If it is only a matter of palliatives, it is because the 'pro-agriculture' talk nurtured by the media in the West is contrasted with an alleged 'preference for industrialization' which is said to be the source of the bankruptcy. For the reason for seeking greater output per farmer is precisely in order to make possible a higher degree of urbanization. But urbanization without industrialization cannot but be parasitical and disastrous. Conversely, industry (but not just any industry) is necessary to make greater agricultural output possible: it has to supply it with machinery and offer it a growing market in return. That is what the option for an autocentred national and popular strategy means. If this option is rejected in favour of a systematic preference for integration into globalization (i.e., the world capitalist economy), talk of giving priority to agriculture becomes hollow and at worst demagogic. The World Bank Berg Report1 has these characteristics, refraining as it does from making a self-criticism of the Bank which supported the policies that led to the present bankruptcy, refraining from making a critique of import-substitution industrialization (which finds its preferred market in the expansion of demand from the middle classes at the expense of the rural and urban popular classes) which it advocated, and refraining from analyzing the implications of its proposal for export-oriented industry. But this derives its 'comparative advantage' from its low wages, which it thus contributes to reproducing. The contradictions of these 'proposals' are moreover obvious: export-oriented industry presupposes low wages and consequently, low food prices, at the very time that the raising of these latter to inspire the peasants to produce more is being recommended.

The capacity of capitalism in the abstract to 'solve the problem of African development' could be discussed ad infinitum. Concrete capitalism, such as it actually exists, that is, globalized, not only has not 'solved' this problem (it even created it) over the last 150 years (or even over the last 400 years since the slave trade), but envisages nothing for the next 50 years. The challenge will therefore only be taken up by the African peoples, the day that the necessary popular alliances enable them to delink their development from the demands of transnationalization.

The undersigned is confident enough that his present work will give immense knowledge to the people regarding the subject and its crux.

— *Author*

CONTENTS

1

The Growth of Commercial Agriculture

Let us glance first of all at the general statistics on grain production in European Russia. The considerable harvest fluctuations render the data for individual periods or for individual years quite useless. It is necessary to take different periods and the data for a whole number of years. We have at our disposal the following data: for the period of the 60s, the data for 1864-1866 (*Military Statistical Abstract*, IV, St. Petersburg, 1871, data of gubernatorial reports). For the 70s, the returns of the Department of Agriculture for the entire decade (*Historico-Statistical Survey of Russian Industry*, Vol. I, St. Petersburg, 1883). And lastly, for the 1880s, we have data for the five years of 1883-1887 (*Statistics of the Russian Empire*, IV); this five-year period can represent the whole of the eighties, since the average harvest for the ten years, 1880-1889, is even somewhat higher than for the five years 1883-1887 (see *Agriculture and Forestry in Russia*, published for the Chicago Exhibition, pp. 132 and 142). Further, in order to judge of the trend of evolution in the 90s we take the data for the decade 1885-1894 (*Productive Forces*, I, 4). Lastly, the data for 1905 (*Yearbook of Russia*, 1906) are quite adequate for a judgement of the present time. The 1905 harvest was only a little lower than the average for the five years 1900-1904.

We see from this that until the 1890s the post-Reform era is characterised by an undoubted increase in the production both of cereals and potatoes. The productivity of agricultural labour rises: firstly, the size of the net yield grows faster than that of the sown area (with occasional exceptions); secondly, we must bear in mind that the proportion of the population engaged in agricultural production steadily diminished during this period owing to the diversion of the population from agriculture to commerce and industry, and also owing to the migration of peasants beyond the bounds of European Russia. What is particularly noteworthy is the fact that it is *commercial* agriculture that is growing: there is an increase in the amount of grain gathered (after subtracting seed) per head of the population, while among this population there is an ever-growing division of social labour; there is an increase in the commercial and industrial population; the agricultural population splits up into rural entrepreneurs and a rural proletariat; there is an extension of specialisation in agriculture itself, so that the amount of grain produced for sale grows far more rapidly than the total amount of grain produced in the country.

The capitalist character of the process is strikingly illustrated by the increased share of potatoes in the total agricultural production. The increase in the area under potatoes signifies, on the one hand, an improvement in agricultural technique (the introduction of root-crops) and increased technical processing of agricultural produce (distilling and the manufacture of potato starch). On the other hand, it is, from the viewpoint of the rural entrepreneur class, the production of relative surplus-value (cheapening of the cost of maintaining labour-power, deterioration of the people's nourishment). The data for the decade 1885-1894 show further that the crisis of 1891-1892, which tremendously intensified the expropriation of the peasantry, led to a considerable reduction in the output of cereals and to a reduction in the yield of all crops; but the process of the displacement of cereals by potatoes continued with such force that the per-capita output of potatoes increased, not withstanding the reduced yield. Finally, the last five years (1900-1904) also show an increase in agricultural production, an increase in the productivity of agricultural labour and a worsening of the conditions of the working-class (increase in the part played by potatoes).

As we have noted above, the growth of commercial agriculture manifests itself in the specialisation of agriculture. Mass-scale and gross data on the production of all crops can give (and then not always) only the most general indications of this process, since the specific features of the different areas thereby disappear. Yet it is precisely the segregation of the different agricultural areas that is one of the most characteristic features of post-Reform agriculture in Russia. Thus, the *Historico-Statistical Survey of Russian Industry* (Vol. I, St. Petersburg, 1883), quoted by us, enumerates the following agricultural areas: the flax area, "the region where stock farming predominates," and where, in particular, "dairy farming is considerably developed"; the region where grain crops predominate, particularly the three-field area and the area with the improved fallow or multi-field grass system (part of the steppe belt, which "is characterised by the production of the most valuable, so-called elite grains, mainly intended for the foreign market"); the beet area, and the area in which potatoes are cultivated for distilling purposes.

The economic areas indicated have arisen in European Russia comparatively recently, and with every passing year continue increasingly to develop and become more segregated" (*loc. cit.*, p. 15). Our task should now be, consequently, to study this process of the specialisation of agriculture, and we should ascertain whether a growth of commercial agriculture is to be observed in its various forms, whether capitalist agriculture comes into existence in the process, and whether agricultural capitalism bears the features we indicated above in analysing the general data on peasant and landlord farming. It goes without saying that for our purposes it will be sufficient if we confine ourselves to describing the principal areas of commercial agriculture.

But before examining the data for the separate areas, let us note the following: the Narodnik economists, as we have seen, do all they can to evade the fact that the characteristic feature of the post-Reform period is the growth of commercial agriculture. Naturally, in doing so they also ignore the circumstance that the drop in grain prices is bound to stimulate the specialisation of agriculture and the drawing of agricultural produce into the sphere of exchange. Here is an instance. The authors of the well-known book *The Influence of Harvests and Grain Prices* all proceed from the postulate that the price of grain is of no importance to natural economy, and they repeat this "truism" endlessly. One of them, Mr. Kablukov, has observed, however, that *under the general conditions of commodity production* this postulate is substantially wrong. "It is possible, of course," he writes, "that the grain placed on the market has cost less to produce than that grown on the consumer's farm, in which case it would appear to be in the interest also of the consuming farm to change from cultivating cereals to other crops" (or to other occupations, we would add), "and, consequently, for it too the market price of grain assumes importance as soon as it fails to coincide with *its* cost of production" (I, 98, note, author's italics). "But we cannot take that into account," he says peremptorily. Why is that? Because, it seems: 1) a change-over to other crops is possible "only where certain conditions exist." By means of this empty truism (everything on earth is possible only under certain conditions!) Mr. Kabłukov calmly evades the fact that the post-Reform period in Russia has created, and continues to create, the very conditions that call for the specialisation of agriculture and the diversion of the population from agriculture. . . . 2) Because "in our climate it is impossible to find a crop equal to cereals in food value". The argument is highly original, expressing a mere evasion of the issue. What has the food value of other crops to do with the matter, if we are dealing with the sale of these other crops and the purchase of cheap grain? . . . 3) Because "grain farms of the consuming type always have a rational basis for their existence." In other words, because Mr. Kablukov "and colleagues" regard natural economy as "rational." The argument, as you see, is irrefutable. . . .

THE COMMERCIAL GRAIN-FARMING AREA

This area covers the outer region in the south and the east of European Russia, the steppe gubernias of Novorossia and the Transvolga. Agriculture is distinguished here for its extensive character and the enormous scale of the production of grain for sale. If we take the eight gubernias of Kherson, Bessarabia, Taurida, Don, Ekaterinoslav, Saratov, Samara and Orenburg, we shall find that in 1883-1887 the net crop of cereals (not including oats) for a population of 13,877,000 amounted to 41.3 million chetverts, i.e., more than one-fourth of the total net yield of the 50 gubernias of European Russia. The crop most commonly sown here is wheat—the principal export grain. Agriculture develops here fastest of all (by comparison with the other areas of Russia), and these gubernias relegate the central black-earth gubernias, formerly in the lead, to the background:

Thus there is a *shifting* of the principal centre of grain production: in the 1860s and 1870s the central black-earth gubernias were ahead of all the rest, but in the 1880s they yielded priority to the steppe and Lower Volga gubernias: their production of grain began to *diminish*.

This interesting fact of the enormous growth of agricultural production in the area described is to be explained by the circumstance that in the post-Reform period the outer steppe regions have been *colonies* of the central, long-settled part of European Russia. The abundance of free land has attracted an enormous stream of settlers, who have quickly increased the area under crops. The extensive development of *commercial* crops was possible only because of the close economic ties of these colonies with central Russia, on the one hand, and the European grain importing countries, on the other. The development of industry in central Russia and the development of commercial farming in the outer regions are inseparably connected and create a market for each other. The industrial gubernias received grain from the South, selling there the products of their factories and supplying the colonies with labour, artisans, and means of production (timber, building materials, tools, etc.). Only because of this social division of labour could the settlers in the steppe localities engage exclusively in agriculture and sell huge quantities of grain in the home and particularly in the foreign market. Only because of their close connection with the home and foreign markets could the economic development of these localities proceed so rapidly; and it was precisely capitalist development, for along with the growth of commercial farming there was an equally rapid process of the diversion of the population into industry, the process of the growth of towns and of the formation of new centres of large-scale industry.

As to the question of whether the growth of commercial farming in this area is bound up with technical progress in agriculture and with the creation of capitalist relations, that has been dealt with above. In Chapter II we saw how large the areas cultivated by peasants in these localities are and how sharply capitalist relations manifest themselves there even within the village community. In the preceding chapter we saw that in this area there has been a particularly rapid development in the use of machinery, that the capitalist farms in the outer regions attract hundreds of thousands and millions of wage-workers, with huge farms created on a scale unprecedented in agriculture, on which there is extensive co-operation of wage-workers, etc. We have little left now to add in completion of this picture.

In the outer steppe regions the privately-owned estates are not only distinguished occasionally for their enormous size, but are also the scene of farming on a very big scale. Above we made reference to crop areas of 8, 10 and 15 thousand dessiatines in Samara Gubernia. In Taurida Gubernia, Falz-Fein owns 200,000 dess., Mordvinov 80,000 dess.; two individuals own 60,000 dess. each, "and many proprietors have from 10,000 to 25,000 dessiatines" (Shakhovskoi, 42). An idea of the scale of farming can be obtained, for example, from the fact that in 1893 there were 1,100 machines (of which 1,000 belonged to the peasantry) haymaking for Falz-Fein. In Kherson Gubernia there were 3.3 million dessiatines under cultivation in 1893, of which 1.3 million dess. belonged to private owners; in five uyezds of the gubernia (without Odessa Uyezd) there were 1,237 medium-sized farms (250 to 1,000 dess. of land), 405 big farms(1,000 to 2,500 dess.) and 226 farms each of over 2,500 dess. According to data gathered in 1890 on 526 farms, they employed 35,514 workers, i.e., an average of 67 workers per farm, of whom from 16 to 30 were annual labourers. In 1893, 100 more or less big farms in Elisavetgrad Uyezd employed 11,197 workers (an average of 112 per farm!), of whom 17.4% were annual, 39.5% seasonal, and 43.1% day labourers. Here are data on the distribution of crop area among all the agricultural undertakings in the uyezd, both of private landowners and of peasants.

Thus, a little over 3 per cent of the peasants (and if we count only those who cultivated, 4 per cent) concentrate in their hands more than a third of the total area under crops, for the tilling and harvesting of which masses of seasonal and day labourers are required.

There is no need, apparently, to comment on these data. We have had occasion to observe that the area described is the most typical of agricultural capitalism in Russia—typical not in the agricultural sense, of course, but in the social-economic sense. These colonies, having developed with the greatest freedom, show us what relations could and should have developed in the rest of Russia, had not the numerous survivals of pre-Reform life retarded the development of capitalism. The forms, however, of agricultural capitalism, as will be seen from what follows, are extremely varied.

THE COMMERCIAL STOCK-FARMING AREA. GENERAL DATA ON THE DEVELOPMENT OF DAIRY FARMING

We now pass to another very important area of agricultural capitalism in Russia, namely, the region in which not cereal, but livestock produce is of predominant significance. This region embraces, apart from the Baltic and the western gubernias, the northern, the industrial and parts of some of the central gubernias (Ryazan, Orel, Tula, and Nizhni Novgorod). Here animals are kept for dairy produce, and the whole character of agriculture is adapted to obtaining as large a quantity as possible of the more valuable market produce of this sort. "Before our very eyes a marked transition is taking place from stock farming for manure to stock farming for dairy produce; it has been particularly notice able during the past ten years" (work quoted in previous footnote, *ibid.*). It is very difficult by the use of statistics to describe the various regions of Russia in this respect, because it is not the total number of horned cattle that is here important, but the number of dairy cattle and their quality. If we take the total number of animals per hundred inhabitants, we shall find that it is biggest in the outer steppe regions of Russia and smallest in the non-black-earth belt (*Agriculture and Forestry*, 274); we shall find that as time goes on the number diminishes (*Productive Forces*, III, 6. Cf. *Historico-Statistical Survey*, I). Hence, we observe here what Roscher noted in his day, namely, that the number of animals per unit of the population is largest in districts of "extensive livestock farming" (W. Roscher, *Nationaloekonomik des Ackerbaues*. 7-te Aufl., Stuttg., 1873, S. 563–564). We, however, are interested in intensive livestock farming, and in dairy farming in particular. We are compelled, therefore, to confine ourselves to the *approximate* computation made by the authors of the above-mentioned, *Sketch*, without claiming to make an exact estimate of the phenomenon; such a computation clearly illustrates the relative positions of the various regions of Russia as to degree of dairy-farm development. We quote this computation in *extenso*, supplementing it with some averages arrived at and data on the cheese-making industry in 1890 according to "factory" statistics.

To judge the development of dairy farming, we can only make use of data on butter production and cheese making. This industry arose in Russia at the very end of the 18th century (1795); cheese making on landlords' estates began to develop in the 19th century, but suffered a severe crisis in the 1860s, which opened the period of cheese making by peasants and merchants.

Thus, in 25 years production increased more than ten-fold; only the dynamics of the phenomenon may be judged from these data, which are extremely incomplete. Let us quote some more detailed material. In Vologda Gubernia an improvement in dairy farming began, properly speaking, in 1872, when the Yaroslavl-Vologda railway was opened; since then "farmers have begun to see to the improvement of their herds, to introduce grass cultivation, to acquire improved implements . . . and have tried to place dairy farming on a purely commercial basis" (*Statistical Sketch*, 20). In Yaroslavl Gubernia "the ground was prepared" by the so-called "cheese-making artels" of the 70s, and "cheese making continues to develop on the basis of private enterprise, merely retaining the title of 'artel'" (25); cheese making "artels" figure—may we add—in the *Directory of Factories and Works* as establishments employing wage-workers. Instead of 295,000 rubles, the authors of the *Sketch* estimate the output of cheese and butter, according to *official* returns, at 412,000 rubles (computed from figures scattered throughout the book); correction of the figure brings the value of the output of fresh butter and cheese to 1,600,000 rubles, and if we add clarified butter and soft cheese, to 4,701,400 rubles, not counting either the Baltic or the western gubernias.

For the later period let us quote the following opinions from the above-cited publication of the Department of Agriculture *Hired Labour, etc*. Concerning the industrial gubernias in general we read: "A complete revolution in the position of the farms in this area has been brought about by the development of dairy farming"; it "indirectly has also helped to bring about an improvement in agriculture"; "dairy farming in the area is developing with every year". In Tver Gubernia "there is to be observed the tendency both among private landowners and peasants to improve the methods of maintaining cattle"; the income from stock farming is estimated at 10 million rubles (274). In Yaroslavl Gubernia "dairy farming . . . is developing with every year. . . . Cheese and butter making have even begun to assume something of an industrial character . . . milk . . . is bought up from neighbours and even from peasants. One comes across cheese factories run by a whole company of owners" (285). "The general trend of private-landowner farming here," writes a correspondent from Danilov Uyezd, Yaroslavl Gubernia, "is marked at the present time by the following: 1) the transition from three-field to five- and seven-field crop rotation, with the sowing of herbage in the fields; 2) the ploughing up of disused lands; 3) the introduction of dairy farming, and as a consequence, the stricter selection of cattle and an improvement in their maintenance" (292). The same thing is said of Smolensk Gubernia, where the value of the output of cheese and butter amounted to 240,000 rubles in 1889—according to a report of the Governor (according to statistical returns, 136,000 rubles in 1890). The development of dairy farming is noted in the Kaluga, Kovno, Nizhni-Novgorod, Pskov, Esthland and Vologda gubernias. The value of the output of butter and cheese in the last-mentioned gubernia was estimated at 35,000 rubles according to statistics for 1890, to 108,000 rubles according to the Governor's report, *and to 500,000 rubles* according to local returns for 1894, which gave a total of *389 factories*. "That is what the statistics say. Actually, however, there are far more factories, since, according to investigations by the Vologda Zemstvo Board, there are 224 factories in Vologda Uyezd alone." Production is developed in three uyezds, and has partly penetrated a fourth. One can judge from this how many times the above quoted figures need to be multiplied in order to approach the real situation. The plain view of an expert that at the present time the number of butter and cheese-making establishments "amounts to several thousand" (*Agriculture and Forestry in Russia*, 299), gives a truer picture of the facts than the allegedly exact figure of 265.

Thus the data leave not the slightest doubt about the enormous development of this special type of commercial farming. The growth of capitalism was accompanied here too by the transformation of routine technique. "In the sphere of cheese making," we read, for example, in *Agriculture and Forestry*, "more has been done in Russia during the last 25 years than perhaps in any other country" (301). Mr. Blazhin says the same thing in his article "Technical Progress in Dairy Farming" (*Productive Forces*, III, 38-45). The principal change is that the "age-old" method of leaving cream to settle has been replaced by the system of separating cream in centrifugal machines (separators). The machine has enabled the work to be carried on irrespective of atmospheric temperature, increased the butter yield from milk by 10%, improved the quality of the product, reduced the cost of butter production (the machine requires less labour, space, and ice, as well as fewer utensils), and has led to the concentration of production. Large peasant butteries have grown up, handling "as much as 500 poods of milk a day, which was physically impossible . . . when the milk was left to settle" (*ibid.*). Improvements are being made in the instruments of production (permanent boilers, screw presses, improved cellars), and production is being assisted by bacteriology, which is providing pure cultures of the type of lactic-acid bacilli needed for fermenting cream.

Thus, in the two areas of commercial farming we have described, the technical improvements called into being by the requirements of the market were effected primarily in those operations that were easiest to change and are particularly important for the market: reaping, threshing and winnowing in commercial grain farming, and the technical processing of animal produce in the area of commercial stock farming. As to the keeping of cattle, capital finds it more profitable for the time being to leave that to the small producer: let him "diligently" and "industriously" tend "his" cattle (and charm Mr. V. V. with his diligence—see *Progressive Trends*, p. 73), let him bear the brunt of the hardest and roughest work of tending the milk-yielding machine. Capital possesses the latest improvements and methods not only of separating the cream from the milk, but also of separating the "cream" from this "diligence", of separating the milk from the children of the peasant poor.

CONTINUATION. THE ECONOMY OF LANDLORD FARMING IN THE AREA DESCRIBED

We have cited the evidence of agronomists and farmers to the effect that dairy farming on the landlord estates leads to the rationalisation of agriculture. Let us add here that the analysis of the Zemstvo statistics on this question made by Mr. Raspopin fully confirms this conclusion. We refer the reader to Mr. Raspopin's article for detailed data and give here only his main conclusion. "The interdependence of the condition of stock raising and dairy farming, on the one hand, and the number of dilapidated estates and the intensity of farming, on the other, is beyond question. The uyezds (of Moscow Gubernia) where dairy cattle raising, dairy farming, is most developed show the smallest percentage of dilapidated farms and the highest percentage of estates with highly developed field cultivation. Throughout Moscow Gubernia ploughland is being reduced and turned into meadow and pastureland, while grain rotations are yielding place to multi-field herbage rotations. Fodder grasses and dairy cattle, and not grain, are now predominant . . . not only on the farming estates in Moscow Gubernia but throughout the Moscow industrial district" (*loc. cit.*).

The scale of butter production and cheese making is particularly important precisely because it testifies to a complete revolution in agriculture, which becomes entrepreneur farming and breaks with routine. Capitalism subordinates to itself one of the products of agriculture, and all other aspects of farming are fitted in with this principal product. The keeping of dairy cattle calls forth the cultivation of grasses, the change-over from the three-field system to multi-field systems, etc. The waste products of cheese making go to fatten cattle for the market. Not only milk processing, but the whole of agriculture becomes a commercial enterprise. The influence of cheese production and butter making is not confined to the farms on which they are carried on, since milk is often bought up from the surrounding peasants and landlords. By buying up the milk, capital subordinates to itself the small agriculturists too, particularly with the organisation of the so-called "amalgamated dairies," the spread of which was noted in the 70s.

These are establishments organised in big towns, or in their vicinity, which process very large quantities of milk brought in by rail. As soon as the milk arrives the cream is skimmed and sold fresh, while the skimmed milk is sold at a low price to poorer purchasers. To ensure that they get produce of a certain quality, these establishments sometimes conclude contracts with the suppliers, obliging them to adhere to certain rules in feeding their cows. One can easily see how great is the significance of large establishments of this kind: on the one hand they capture the public market (the sale of skimmed milk to the poorer town-dwellers), and on the other hand they enormously expand the market for the rural entrepreneurs. The latter are given a tremendous impetus to expand and improve commercial farming. Large-scale industry brings them into line, as it were, by demanding produce of a definite quality and forcing out of the market (or placing at the mercy of the usurers) the small producer who falls below the "normal" standard. There should also operate in the same direction the grading of milk as to quality (fat content, for example), on which technicians are so busily engaged, inventing all sorts of lacto-densimeters, etc., and of which the experts are so heartily in favour (cf. *Productive Forces*, III, 9 and 38).

In this respect the role of the amalgamated dairies in the development of capitalism is quite analogous to that of elevators in commercial grain farming. By sorting grain as to quality the elevators turn it into a product that is not individual but generic (*res fungibilis*, as the lawyers say), i.e., for the first time they adapt it fully to exchange (cf. M. Sering's article on the grain trade in the United States of America in the symposium *Landownership and Agriculture*, p. 281 and foll.). Thus, the elevators give a powerful impetus to commodity-grain production and spur on its technical development by also introducing grading for quality. Such a system strikes a double blow at the small producer. Firstly, it sets up as a standard, legalises, the higher-quality grain of the big crop sowers and thereby greatly depreciates the inferior grain of the peasant poor. Secondly, by organising the grading and storing of grain on the lines of large-scale capitalist industry, it reduces the big sowers' expenses on this item and facilitates and simplifies the sale of grain for them, thereby placing the small producer, with his patriarchal and primitive methods of selling from the cart in the market, totally at the mercy of the kulaks and the usurers. Hence, the rapid development of elevator construction in recent years means as big a victory for capital and degradation of the small commodity-producer in the grain business as does the appearance and development of capitalist "amalgamated dairies."

From the foregoing material it is clear that the development of commercial stock farming *creates* a home market, firstly, for means of production—milk-processing equipment, premises, cattle sheds, improved agricultural implements required for the change-over from the routine three-field system to multi-field crop rotations, etc.; and secondly, for labour-power. Stock farming placed on an industrial footing requires a far larger number of workers than the old stock farming "for manure." The dairy farming area—the industrial and north-western gubernias—does really attract masses of agricultural labourers. Very many people go to seek agricultural work in the Moscow, St. Petersburg, Yaroslavl and Vladimir gubernias; fewer, but nevertheless a considerable number, go to the Novgorod, Nizhni Novgorod and other non-black-earth gubernias. According to correspondents of the Department of Agriculture in the Moscow and other gubernias private-landowner farming is actually conducted in the main by labourers from other areas. This paradox— the migration of agricultural workers from the agricultural gubernias (they come mostly from the central black-earth gubernias and partly from the northern) to the industrial gubernias to do agricultural jobs in place of industrial workers who abandon the area en masse—is an extremely characteristic phenomenon.

It proves more convincingly than do any calculations or arguments that the standard of living and the conditions of the working people in the central black-earth gubernias, the least capitalist ones, are incomparably lower and worse than in the industrial gubernias, the most capitalist ones; it proves that in Russia, too, the following has become a universal fact, namely, the phenomenon characteristic of all capitalist countries, that the conditions of the workers in industry are better than those of the workers in agriculture (because in agriculture oppression by capitalism is supplemented by the oppression of pre-capitalist forms of exploitation). That explains the flight from agriculture to industry, whereas not only is there no flow from the industrial gubernias towards agriculture (for example, there is no migration from these gubernias at all), but there is even a tendency to look down upon the "raw" rural workers, who are called "cowherds" (Yaroslavl Gubernia), "cossacks" (Vladimir Gubernia) and "land labourers" (Moscow Gubernia).

It is important also to note that cattle herding requires a larger number of workers in winter than in summer. For that reason, and also because of the development of agricultural processing trades, the demand for labour in the area described not only grows, but is *more evenly distributed over the whole year* and over a period of years. The most reliable material for judging this interesting fact is the data on wages, if taken for a number of years. We give these data, confining ourselves to the groups of Great-Russian and Little-Russian gubernias. We omit the western gubernias, owing to their specific social conditions and artificial congestion of population (the Jewish pale of settlement), and quote the Baltic gubernias only to illustrate the relations that arise where capitalism is most highly developed.

Let us examine this table, in which the three principal columns are printed in italics. The first column shows the proportion of summer to yearly pay. The *lower* this proportion is, and the nearer the summer pay approximates to half the yearly pay, the more evenly is the demand for labour spread over the entire year and *the less the winter unemployment*. The least favourably placed in this respect are the central black-earth gubernias—the area where labour-service prevails and where capitalism is poorly developed. In the industrial gubernias, in the dairy-farming area the demand for labour is higher and winter unemployment is less. Over a period of years, too, the pay is most stable here, as may be seen from the second column, which shows the difference between the lowest and the highest pay in the harvest season. Lastly, the difference between the pay in the sowing season and the pay in the harvest season is also least in the non black-earth belt, i.e., the demand for workers is more evenly distributed over the spring and summer. In all respects mentioned the Baltic gubernias stand even higher than the non-black-earth gubernias, while the steppe gubernias, with their immigrant workers and with harvest fluctuations of the greatest intensity, are marked by the greatest instability of wages. Thus, the data on wages testify that agricultural capitalism in the area described not only creates a demand for wage-labour, but also distributes this demand more evenly over the whole year.

Lastly, reference must be made to one more type of dependence of the small agriculturist in the area described upon the big farmer. This is the replenishment of landlords' herds by the purchase of cattle from peasants. The landlords find it more profitable to buy cattle from peasants driven by need to sell "at a loss" than to breed cattle themselves—just as our buyers-up in so-called handicraft industry often prefer to buy finished articles from the handicraftsmen at a ruinously cheap price rather than manufacture them in their own workshops. This fact, which testifies to the extreme degradation of the small producer, and to his being able to keep going in modern society only by endlessly reducing his requirements, is turned by Mr. V. V. into an argument in favour of small "people's" production! . . . "We are entitled to draw the conclusion that our big farmers . . . do not display a sufficient degree of independence. . . . The peasant, however . . . reveals greater ability to effect real farming improvements" (*Progressive Trends*, 77). This lack of independence is expressed in the fact that "our dairy farmers . . . buy up the peasants' (cows) at a price rarely amounting to half the cost of raising them—usually at not more than a third, and often even a quarter of this cost" (*ibid.*, 71). The merchant's capital of the stock farmers has made the small peasants completely dependent, it has turned them into its cowherds, who breed cattle for a mere song, and has turned their wives into its milkmaids. One would think that the conclusion to be drawn from this is that there is no sense in retarding the transformation of merchant's capital into industrial capital, no sense in supporting small production, which leads to forcing down the producer's standard of living below that of the farm labourer. But Mr. V. V. thinks otherwise. He is delighted with the "zeal" (p. 73, *loc. cit.*) of the peasant in tending his cattle; he is delighted with the "good results from livestock farming" obtained by the peasant woman who "spends all her life with her cow and sheep" (80). What a blessing, to be sure! To "spend all her life with her cow" (the milk of which goes to the improved cream separator), and as a reward for this life, to receive "one-fourth of the cost" of tending this cow! Now really, how after that can one fail to declare in favour of "small people's production"!

CONTINUATION. THE DIFFERENTIATION OF THE PEASANTRY IN THE DAIRY-FARMING AREA

In the literature dealing with the effect of dairy farming on the conditions of the peasantry, we constantly come up against contradictions: on the one hand reference is made to progress in farming, the enlargement of incomes, the improvement of agricultural technique and the acquisition of improved implements; on the other hand, we have statements about the deterioration of food, the creation of new types of bondage and the ruin of the peasants. After what was stated in Chapter II, we should not be surprised at these contradictions: we know that these opposite opinions relate to opposite groups of the peasantry. For a more precise judgement of the subject, let us take the data showing the classification of peasant households according to the number of cows per household.

Thus, the distribution of cows among the peasants in the non-black-earth belt is found to be very similar to the distribution of draught animals among the peasants in the black-earth gubernias (see Chapter II). Moreover, the concentration of dairy cattle in the area described proves to be greater than the concentration of draught animals. This clearly points to the fact that it is with the local form of commercial farming that the differentiation of the peasantry is closely connected. The same connection is evidently indicated by the following data (unfortunately, not sufficiently complete). If we take the aggregate Zemstvo statistics (given by Mr. Blagoveshchensky; for 122 uyezds of 21 gubernias), we get an average of 1.2 cows per household. Hence, in the non-black-earth belt the peasantry evidently own more cows than in the black-earth belt, and in Petersburg Gubernia they are better off than in the non-black earth belt in general. On the other hand, in 123 uyezds of 22 gubernias the cattleless households constitute 13 %, while in the 18 uyezds we have taken, they amount to 17%, and in the 6 uyezds of Petersburg Gubernia 18.8%. Hence, the differentiation of the peasantry (in the respect we are now examining) is most marked in Petersburg Gubernia, followed by the non-black-earth belt in general. By this indication, commercial farming is the principal factor in the differentiation of the peasantry.

The data show that about half the peasant households (those having no cows, or one cow) can take only a negative part in the benefits of dairy farming. The peasant with one cow will sell milk only out of need, to the detriment of his children's nourishment. On the other hand, about one-fifth of the households (those with 3 cows and more) concentrate in their hands probably more than half the total dairy farming since the quality of their cattle and the profitableness of their farms should be higher than in the case of the "average" peasant. An interesting illustration of this conclusion is provided by the data on a locality where dairy farming and capitalism in general are highly developed. We refer to Petersburg Uyezd. Dairy farming is particularly widely developed in the summer residential part of the uyezd, inhabited mainly by Russians; here the most widely cultivated crops are: grasses (23.5% of the allotment arable, as against 13.7% for the uyezd), oats (52.3% of the arable) and potatoes (10.1%). Agriculture is directly influenced by the St. Petersburg market, which needs oats, potatoes, hay, milk and horse traction (*loc. cit.*, 168). The families of the registered population are 46.3% engaged "in the milk industry." Of the total number of cows 91% provide milk for the market. The income from this industry amounts to 713,470 rubles (203 rubles per family, 77 rubles per cow). The nearer the locality is to St. Petersburg, the higher is the quality of the cattle and the better the attention they receive. The milk is sold in two ways: 1) to buyers-up on the spot and 2) in St. Petersburg to "dairy farms," etc. The latter type of marketing is much more profitable, but "the majority of the farms having one or two cows, and sometimes more, are not . . . able to deliver their milk to St. Petersburg direct"—they have no horses, it does not pay to cart small quantities, etc. The buyers-up of the milk include not only specialist merchants, but individuals with dairies of their own.

One can judge from this how the benefits of dairy farming are distributed among all the peasants in the non-black earth belt, among whom, as we have seen, the concentration of dairy cattle is even greater than among these 560 families. It remains for us to add that 23.1% of the peasant families in St. Petersburg Uyezd hire workers (most of whom, here, as everywhere in agriculture, are day labourers). "Bearing in mind that agricultural workers are hired almost exclusively by families having fully-operating farms" (constituting only 40.4% of the total number of families in the uyezd) "the conclusion must be that more than half of such farms do not manage without hired labour" (158).

Thus, at opposite ends of Russia, in the most varying localities, in St. Petersburg and, say, Taurida gubernias, the social and economic relations within the "village community," prove to be absolutely identical. The "muzhik-cultivators" (Mr. N.–on's term) in both places differentiate into a minority of rural entrepreneurs and a mass of rural proletarians. The specific feature of agriculture is that capitalism subjugates one aspect of rural economy in one district, and another aspect in another, which is why identical economic relations are manifested in the most varied forms of agronomy and everyday life.

Having established the fact that in the area described, too, the peasantry splits up into opposite classes, we shall easily achieve clarity about the contradictory opinions usually expressed as to the role of dairy farming. Quite naturally, the well-to-do peasantry receive an incentive to develop and improve their farming methods and as a result grass cultivation is widespread and becomes an essential part of commercial stock-farming. The development of grass cultivation is observed, for example, in Tver Gubernia; in Kashin Uyezd, the most progressive in that gubernia, as many as one-sixth of all peasant households plant clover (*Returns*, XIII, 2, p. 171). It is interesting, moreover, to note that on the purchased lands a larger proportion of arable is occupied by herbage than on the allotments: the peasant bourgeoisie naturally prefer private ownership of land to communal tenure. In the *Survey of Yaroslavl Gubernia* (Vol. II, 1896) we also find numerous references to the increase in grass cultivation, and again mainly on purchased and rented lands. In the same publication we find references to the spread of improved implements: iron ploughs, threshing machines, rollers, etc. Butter and cheese making, etc., are developing very considerably. In Novgorod Gubernia it was noted as far back as the beginning of the 80s that along with a general deterioration and diminution of peasant stock breeding, there was an improvement in certain individual localities where there was a profitable market for milk and where the milk-feeding of calves was an old-established industry (Bychkov: *An Essay in the House-to-House Investigation of the Economic Position and Farming of the Peasants in Three Volosts of Novgorod Uyezd*, Novgorod, 1882). The milk-feeding of calves, which is also a type of commercial livestock farming, is, generally speaking, a fairly widespread industry in the Novgorod and Tver gubernias and in other places not far from the big cities (see *Hired Labour, etc.*, published by the Department of Agriculture). "This industry," says Mr. Bychkov, "by its very nature, brings an income to the already well-provided peasants possessing considerable numbers of cows, since with one cow, and sometimes even with two of poor yield, the milk-feeding of calves is unthinkable" (*loc. cit.*, 101).

But the most outstanding index of the economic successes of the peasant bourgeoisie in the area described is the hiring of labourers by peasants. The local landowners feel that they are being confronted by competitors, and in their communications to the Department of Agriculture they sometimes even attribute the shortage of workers to the fact that these are snatched up by the well-to-do peasants (*Hired Labour*, 490). The hiring of labourers by peasants is noted in the Yaroslavl, Vladimir, St. Petersburg and Novgorod gubernias (*loc. cit., passim*). A mass of such references is also scattered throughout the *Survey of Yaroslavl Gubernia*. This progress of the well-to-do minority, however, is a heavy burden upon the mass of the poor peasants. In Koprin Volost, Rybinsk Uyezd, Yaroslavl Gubernia, for example, one finds the spread of cheese making—on the initiative of "V. I. Blandov, the well-known founder of cheese-making artels " "When the poorer peasants, with only one cow each, deliver . . . their milk (to the cheese factory) they do so, of course, to the detriment of their own nourishment"; whereas the well-to-do peasants improve their cattle (pp. 32-33). Among the types of wage-labour undertaken, one finds employment away from home, at cheese-making establishments; from among the young peasants a body of skilled cheese makers is arising. In the Poshekhonye Uyezd "the number . . . of cheese and butter establishments is increasing from year to year," but "the benefits accruing to peasant farming from cheese and butter making hardly compensate for the disadvantages to peasant life resulting from our cheese and butter establishments."

On the peasants' own admission they are often compelled to starve, for with the opening of a cheese or butter factory in some locality, the milk is sent there and the peasants usually drink diluted milk. The system of payment in kind is coming into vogue (pp.43, 54, 59 and others), so that it is to be regretted that our "people's" petty production is not covered by the law prohibiting payment in kind in "capitalist" factories.

Thus, the opinions of people directly acquainted with the matter confirm our conclusion that the majority of the peasants play a purely negative part in the progress of local agriculture. The progress of commercial farming worsens the position of the bottom groups of peasants and forces them out of the ranks of the cultivators altogether.

Be it noted that reference has been made in Narodnik literature to this contradiction between the progress of dairy farming and the deterioration of the peasants' nourishment (for the first time, I think, by Engelhardt). But it is precisely this example that enables one to see the narrowness of the Narodnik appraisal of the phenomena occurring among the peasantry and in agriculture.

They note a contradiction in one form, in one locality, and do not realise that it is typical of the entire social and economic system, manifesting itself everywhere in different forms. They note the contradictory significance of one "profitable industry," and strongly urge the "implanting" among the peasantry of all sorts of other "local industries." They note the contradictory significance of one form of agricultural progress and do not understand that machines, for example, have exactly the same political and economic significance in agriculture as in industry.

THE FLAX-GROWING AREA

We have described the first two areas of capitalist agriculture in fairly great detail because of their widespread character and of the typical nature of the relations observed there. In our further exposition we shall confine ourselves to briefer remarks on some highly important areas.

Flax is the chief of the so-called "industrial crops." The very term indicates that we are dealing here with commercial farming. For example, in the "flax" gubernia of Pskov, flax has long been the peasants' "first money," to use a local expression (*Military Statistical Abstract*, 260). Flax growing is simply a means of making money.

The post-Reform period is marked on the whole by an undoubted increase in commercial flax growing. Thus, at the end of the 60s, the output of flax in Russia was estimated at approximately 12 million poods of fibre (*ibid.*, 260); at the beginning of the 80s at 20 million poods of fibre (*Historico-Statistical Survey of Russian Industry*, Vol. I, St. Petersburg, 1883, p. 74); at the present time, in the 50 gubernias of European Russia over 26 million poods of fibre are gathered. In the flax-growing area proper (19 gubernias of the non-black-earth belt) the area under flax has changed in recent years as follows; 1893—756,600 dess.; 1894—816,500 dess.; 1895—901,800 dess.; 1896—952,100 dess., and 1897—967,500 dess. For the whole of European Russia (50 gubernias) the figure for 1896 was 1,617,000 dess. under flax and for 1897—1,669,000 dess. (*Vestnik Finansov, ibid.*, and 1898, No. 7), as against 1,399,000 dess. at the beginning of the 1890s (*Productive Forces*, I, 36). Similarly, general opinions expressed in publications also testify to the growth of commercial flax growing.

Thus, regarding the first two decades after the Reform, the *Historico-Statistical Survey* states that "the region of flax cultivation for industrial purposes has been enlarged by several gubernias" (*loc. cit.*, 71), which is due particularly to the extension of the railways. Concerning the Yuryev Uyezd, Vladimir Gubernia, Mr. V. Prugavin wrote at the beginning of the eighties: "The cultivation of flax . . . has become very widespread here during the past 10 to 15 years." "Some large family households sell flax to the extent of 300 to 500 rubles and more per annum. . . . They buy" (flax seed) "in Rostov. . . . The peasants in these parts are very careful in selecting seed" (*The Village Community, Handicraft Industries and Agriculture of Yuryev Uyezd, Vladimir Gubernia,* Moscow, 1884, pp. 86-89). *The Zemstvo Statistical Returns for Tver Gubernia* (Vol. XIII, Pt. 2) notes that "the most important spring grain crops, barley and oats, are yielding place to potatoes and flax" (p. 151); in some uyezds flax occupies from 1/3 to 3/4 of the area under spring crops, for example, in Zubtsov, Kashin and other uyezds, "in which flax growing has assumed the clearly expressed speculative character of an industry" (p. 145), developing particularly on rented virgin and disused land.

Moreover, it is noted that in some gubernias, where free land is still available (virgin soil, wasteland, forest-cleared tracts), flax growing is particularly expanding, but in some of the old established flax-growing gubernias "the cultivation of flax is either on the old scale or is even yielding place, for example, to the newly-introduced cultivation of root-crops, vegetables, etc." (*Vestnik Finansov,* 1898, No. 6, p. 376, and 1897, No. 29), i.e., to other types of commercial farming.

As for flax exports, during the first two decades after the Reform they increased with remarkable rapidity: from an average of 4.6 million poods in the years 1857-1861 to 8.5 million poods in the years 1867-1871 and to 12.4 million poods in the years 1877-1881; but then exports seemed to become stationary, amounting in the years 1894-1897 to an average of 13.3 million poods. The development of commercial flax growing led, naturally, to exchange not only between agriculture and industry (sale of flax and purchase of manufactured goods), but between different types of commercial agriculture (sale of flax and purchase of grain) The following data concerning this interesting phenomenon clearly demonstrate that a home market for capitalism is created not only by the diversion of population from agriculture to industry, but also by the specialisation of commercial farming.

How does this growth of commercial flax growing affect the peasantry, who, as we know, are the principal flax producer? "Travelling through Pskov Gubernia and observing its economic life, one cannot help noticing that side by side with occasional large and rich units, hamlets and villages, there are extremely poor units; *these extremes are a characteristic feature of the economic life of the flax area*." "Flax growing has taken a speculative turn," and "the greater part" of the income from flax "is pocketed by buyers-up and by those who lease out land for flax growing" (Strokin, 22-23). The ruinous rents constitute real "money rent" (see above), and the mass of the peasants are in a state of "complete and hopeless dependence" (Strokin, *ibid.*) upon the buyers-up. The sway of merchant's capital was established in this locality long ago, and what distinguishes the post-Reform period is the enormous concentration of this capital, the undermining of the monopoly of the former small buyers-up and the formation of "flax agencies" which have captured the whole flax trade. The significance of flax growing, says Mr. Strokin about Pskov Gubernia, "is expressed . . . in the concentration of capital in a few hands" (p. 31). Turning flax growing into a gamble, capital ruined vast numbers of small agriculturists, who worsened the quality of the flax, exhausted the land, were reduced to leasing out their allotments and finally swelled the ranks of "migratory" workers. On the other hand, a slight minority of well-to-do peasants and traders were able—and competition made it necessary—to introduce technical improvements. *Couté* scutchers, both hand-worked (costing up to 25 rubles) and horse-operated (three times dearer), were introduced. In 1869 there were only 557 such machines in Pskov Gubernia, in 1881 there were 5,710 (4,521 hand-worked and 1,189 horse- operated). "Today," we read in the *Historico-Statistical Survey*, "every sound peasant family engaged in flax growing has a *Couté* hand-machine, which has actually come to be called the 'Pskov scutcher'" (*loc. cit.*, 82-83). What proportion this minority of "sound" householders who acquire machines is to the rest of the peasantry.

Instead of the primitive contrivances which cleaned the seeds very badly, the Pskov Zemstvo began to introduce improved seed-cleaners (*trieurs*), and "the more prosperous peasant industrialists" now find it profitable to buy these machines themselves and to hire them out to flax growers (*Vestnik Finansov*, 1897, No. 29, p. 85). The bigger buyers-up of flax establish drying rooms and presses and hire workers to sort and scutch the flax (see example given by Mr. V. Prugavin, *loc. cit.*, 115). Lastly, it should be added that the processing of flax-fibre requires quite a large number of workers: it is estimated that the cultivation of one dessiatine of flax requires 26 working days of agricultural work proper, and 77 days to extract the fibre from stalks (*Historico-Statistical Survey*, 72). Thus, the development of flax growing leads, on the one hand, to the farmer being more fully occupied during the winter and, on the other, to the creation of a demand for wage-labour on the part of those landlords and well-to-do peasants who engage in flax growing (see the example in Chapter III, § VI).

Thus, in the flax-growing area, too, the growth of commercial farming leads to the domination of capital and to the differentiation of the peasantry. A tremendous obstacle to the latter process is undoubtedly the ruinously high renting prices of land, the pressure of merchant's capital, the tying of the peasant to his allotment and the high payments for the allotted land. Hence, the wider the development of land purchase by the peasants, and of migration in search of employment, and the more widespread the use of improved implements and methods of cultivation, the more rapidly will merchant's capital be supplanted by industrial capital, and the more rapidly will a rural bourgeoisie be formed from among the peasantry, and the system of labour-service for the landlord replaced by the capitalist system.

THE TECHNICAL PROCESSING OF AGRICULTURAL PRODUCE

Above we have already had occasion to note (Chapter I, § I) that writers on agriculture, in classifying systems of farming according to the principal market product, assign the industrial or technical system of farming to a special category. The essence of this system is that the agricultural product, before going into consumption (personal or productive), undergoes technical processing. The establishments which effect this processing either constitute part of the very farms on which the raw material is produced or belong to special industrialists who buy up the raw material from the peasant farmers. From the standpoint of political economy the difference between these two types is unimportant. The growth of agricultural technical trades is extremely important as regards the development of capitalism. Firstly, this growth represents one of the forms of the development of commercial farming, and is, moreover, the form that shows most vividly the conversion of agriculture into a branch of industry of capitalist society.

Secondly, the development of the technical processing of agricultural produce is usually connected intimately with technical progress in agriculture: on the one hand, the very production of the raw material for processing often necessitates agricultural improvement (the planting of root-crops, for example); on the other hand, the waste products of the processing are frequently utilised in agriculture, thus increasing its effectiveness and restoring, at least in some measure, the equilibrium, the interdependence, between agriculture and industry, the disturbance of which constitutes one of the most profound contradictions of capitalism.

We must accordingly now describe the development of technical agricultural trades in post-Reform Russia.

1) Distilling

Here we regard distilling only from the point of view of agriculture. Accordingly, there is no need for us to dwell on the rapid concentration of distilling in large plants (partly due to excise requirements), on the rapid progress of factory technique, with the consequent cheapening of production, and the increase in excise duties which has outstripped this cheapening of production and because of its excessive amount has retarded the growth of consumption and production.

Thus, over 9/10 of the distilleries (accounting for over 4/5 of the total output) are directly connected with agriculture. Being large capitalist enterprises, these establishments lend the same character to all the landlord farms on which they are set up (the distilleries belong almost without exception to landlords, mainly to members of the nobility). The type of commercial farming under review is particularly developed in the central black-earth gubernias, in which are concentrated over 1/10 of the total number of distilleries in the Russian Empire (239 in 1896-97, of which 225 were agricultural and mixed), producing over a quarter of the total output of spirits (7,785,000 vedros in 1896-97, of which 6,828,000 at agricultural and at mixed distilleries). Thus in the area where labour-service predominates, the commercial character of agriculture most frequently (as compared with other areas) manifests itself in the distilling of vodka from grain and potatoes. Distilling from potatoes has undergone a particularly rapid development since the Reform, as may be seen from the following data relating to the whole of the Russian Empire:

Thus, with a general twofold increase in the quantity of crops distilled, the quantity of potatoes used increased about 15-fold. This fact strikingly corroborates the proposition established above (§ I in this chapter) that the enormous increase in the potato area and crop signifies the growth of precisely commercial and capitalist farming, along with improvement of agricultural technique, with the replacement of the three-field system by multi-field crop rotation, etc. The area of the biggest development of distilling is also distinguished for the biggest (in the Russian gubernias, i.e., not counting the Baltic and the western gubernias) net per-capita harvest of potatoes. Thus in the northern black-earth gubernias the figures for 1864-1866, 1870-1879 and 1883-1887 were 0.44, 0.62 and 0.60 chetverts respectively, whereas for the whole of European Russia (50 gubernias) the corresponding figures were 0.27, 0.43 and 0.44 chetverts. As far back as the beginning of the 80s the *Historico-Statistical Survey* noted that "the region marked by the greatest expansion of potato cultivation covers all the gubernias of the central and northern parts of the black-earth belt, the Volga and Transvolga gubernias and the central non-black-earth gubernias" (*loc. cit.*, p. 44).

The expansion of potato cultivation by landlords and well-to-do peasants means an increase in the demand for hired labour; the cultivation of a dessiatine of potatoes absorbs much more labour than the cultivation of a dessiatine of cereals and the use of machinery in, for example, the central black-earth area is still very slight. Thus, while the number of workers engaged in the distilling industry proper has decreased, the elimination of labour-service by the capitalist system of farming, with the cultivation of root-crops, has increased the demand for rural day labourers.

2) Beet-Sugar Production

The processing of sugar-beet is even more highly concentrated in big capitalist enterprises than distilling is, and is likewise an adjunct of the landlords' (mainly noblemen's) estates. The principal area of this industry is the south-western gubernias, and then the southern black-earth and central black-earth gubernias. The area under sugar-beet amounted in the 60s to about 100,000 dess., in the 70s to about 160,000 dess.; in 1886-1895 to 239,000 dess., in 1896-1898 to 369,000 dess., in 1900 to 478,778 dess., in 1901 to 528,076 dess. (*Torgovo-Promyshlennaya Gazeta*, 1901, No. 123), in 1905-06 to 483,272 dess. (*Vestnik Finansov*, 1906, No. 12). Hence, in the period following the Reform the area cultivated has increased more than 5-fold. Incomparably more rapid has been the growth of the amount of sugar-beet harvested and processed: on an average the weight of sugar-beet processed in the Empire in the years 1860-1864 was 4.1 million berkovets; in 1870-1874—9.3 million; in 1875-1879—12.8 million; in 1890-1894—29.3 million; and in 1895-96 and 1897-98—35 million. The amount of processed sugar-beet has grown since the 60s more than 8-fold. Hence, there has been an enormous increase in the beet yield, i.e., in labour productivity, on the big estates organised on capitalist lines. The introduction of a root-plant like beet into the rotation is indissolubly linked with the transition to a more advanced system of cultivation, with improved tillage and cattle feed, etc. "The tillage of the soil for beetroot," we read in the *Historico-Statistical Survey* (Vol. I), "which, generally speaking, is rather complicated and difficult, has been brought to a high degree of perfection on many beet farms, especially in the south-western and Vistula gubernias. In different localities, various more or less improved implements and ploughs are used for tilling; in some cases even steam ploughing has been introduced" (p. 109).

This progress of large-scale capitalist farming gives rise to quite a considerable increase in the demand for agricultural wage-workers— regular and particularly day labourers—the employment of female and child labour being particularly extensive (cf. *Historico-Statistical Survey*, II, 32). Among the peasants of the neighbouring gubernias a special type of migration has arisen, known as migration "to sugar" (*ibid.*, 42). It is estimated that the complete cultivation of a morg (= 2/3 dess.) of beet land requires 40 working days (*Hired Labour*, 72). The *Combined Material on the Position of the Rural Population* (published by Committee of Ministers) estimates that the cultivation of one dessiatine of beet land, when done by machine, requires 12, and when by hand 25, working days of males, not counting women and juveniles (pp. X-XI). Thus, the cultivation of the total beet area in Russia probably engages not less than 300,000 agricultural day labourers, men and women. But the increase in the number of dessiatines under beet is not enough to give a complete idea of the demand for hired labour, since some jobs are paid for at so much per berkovets. Here, for example, is what we read in *Reports and Investigations of Handicraft Industry in Russia* (published by Ministry of State Properties, Vol. II, St. Petersburg, 1894, p. 82).

"The female population both of the town, and of the uyezd" (the town of Krolevets, Chernigov Gubernia, is referred to) "think highly of work on the beet fields; in the autumn the cleaning of beets is paid at 10 kopeks per berkovets, and two women clean from six to ten berkovets a day, but some contract to work during the growing season as well, weeding and hoeing; in that case, for the full job, including digging and cleaning, they get 25 kopeks per berkovets of cleaned beets." The conditions of the workers on the beet plantations are extremely bad. For instance, the *Vrachebnaya Khronika Kharkovskoi Gubernii* (September 1899, quoted in *Russkiye Vedomosti*, 1899, No. 254) cites "a number of exceedingly deplorable facts about the conditions *of those working on the red-beet plantations*. Thus, the Zemstvo physician, Dr. Podolsky, of the village of Kotelva, Akhtyrka Uyezd, writes: 'In the autumn *typhus* usually breaks out among *young people employed on the red-beet plantations of the well-to-do peasants*.

The sheds assigned for the workers' leisure and sleeping quarters are kept by such planters in a very filthy condition; by the time the job ends the straw used for sleeping is literally converted into dung, for it is never changed: this becomes a breeding ground of infection. Typhus has had to be diagnosed immediately in the case of four or five patients brought in from one and the same plantation.' In the opinion of this doctor, 'most of the syphilis cases come from the red-beet plantations.' Mr. Feinberg rightly asserts that 'work on the plantations, which is no less injurious to the workers themselves and to the surrounding population than work in the factories, has particularly disastrous consequences, because large numbers of women and juveniles are engaged in it, and because the workers here are without the most elementary protection from society and the State'; in view of this, the author wholly supports the opinion expressed by Dr. Romanenko at the Seventh Congress of Doctors of Kharkov Gubernia that 'in issuing *compulsory* regulations, *consideration* must also be given *to the conditions of the workers on the beet plantations*. These workers lack the most essential things; they live for months under the open sky and eat from a common bowl.'"

Thus, the growth of beet cultivation has enormously increased the demand for rural workers, converting the neighbouring peasantry into a rural proletariat. The increase in the number of rural workers has been but slightly checked by the inconsiderable drop in the number of workers engaged in the beet-sugar industry proper.

3) Potato-Starch Production

From branches of technical production conducted exclusively on landlord farms let us pass to such as are more or less within the reach of the peasantry. These include, primarily, the processing of potatoes (partly also wheat and other cereals) into starch and treacle. Starch production has developed with particular rapidity in the post-Reform period owing to the enormous growth of the textile industry, which raises a demand for starch. The area covered by this branch of production is mainly the non-black-earth, the industrial, and, partly, the northern black-earth gubernias.

The *Historico-Statistical Survey* (Vol. II) estimates that in the middle of the 60s, there were about 60 establishments with an output valued at about 270,000 rubles, while in 1880 there were 224 establishments with an output valued at 1,317,000 rubles. In 1890, according to the *Directory of Factories and Works* there were 192 establishments employing 3,418 workers, with an output valued at 1,760,000 rubles. "In the past 25 years," we read in the *Historico-Statistical Survey*, "the number of establishments engaged in starch production has increased 4 1/2 times and the total output 10 3/4 times; nevertheless, this productivity is far from covering the demand for starch" (p. 116), as evidenced by the increased starch imports from abroad.

Analysing the data for each gubernia, the *Historico-Statistical Survey* reaches the conclusion that our production of potato starch (unlike that of wheat-starch) is of an agricultural character, being concentrated in the hands of peasants and landlords. "Showing promise of extensive development" in the future, "it is even now furnishing our rural population with considerable advantages" (126).

We shall see in a moment who enjoys these advantages. But first let us note that two processes must be distinguished in the development of starch production: on the one hand, the appearance of new small factories and the growth of peasant production, and on the other, the concentration of production in large steam-powered factories. For instance, in 1890 there were 77 steam-powered factories, with 52% of the total number of workers and 60% of the total output concentrated in them. Of these works only 11 were established before 1870, 17 in the 70s, 45 in the 80s, and 2 in 1890 (Mr. Orlov's *Directory*).

To acquaint ourselves with the economy of peasant starch production, let us turn to local investigations. In Moscow Gubernia, in 1880-81, 43 villages in 4 uyezds engaged in starch making. The number of establishments was estimated at 130, employing 780 workers and having an output valued at not less than 137,000 rubles. The industry spread mainly after the Reform; its technique gradually improved and larger establishments were formed requiring more fixed capital and showing a higher productivity of labour. Hand graters were replaced by improved ones, then horse power appeared, and finally the *drum* was introduced, considerably improving and cheapening production. Here are data we have compiled from a house-to-house census of "handicraftsmen," according to size of establishment: Thus we have here small capitalist establishments in which, as production expands, the employment of hired labour increases and the productivity of labour rises. These establishments bring the peasant bourgeoisie considerable profit, and also improve agricultural technique. But the situation of the workers in these workshops is very unsatisfactory, owing to the extremely insanitary working conditions and the long working day.

The peasants who own "grating" establishments farm under very favourable conditions. The planting of potatoes (on allotment, and chiefly on rented land) yields a considerably larger income than the planting of rye or oats. To enlarge their business the workshop owners rent a considerable amount of allotment land from the poor peasants. For example, in the village of Tsybino (Bronnitsy Uyezd), 18 owners of starch workshops (out of 105 peasant families in the village) rent allotments from peasants who have left in search of employment, and also from horseless peasants, thus adding to their own 61 allotments 133 more, which they have rented; concentrated in their hands are a total of 194 allotments, i.e., 44.5% of the total number of allotments in the village. "Exactly similar things," we read in the *Returns*, "are met with in other villages where the starch industry is more or less developed" (*loc. cit.*, 42). The owners of the starch workshops have twice as much livestock as the other peasants: they average 3.5 horses and 3.4 cows per household, as against 1.5 horses and 1.7 cows among the local peasants in general. Of the 68 workshop owners (covered by the house-to-house census) 10 own purchased land, 22 rent non-allotment land and 23 rent allotment land. In short, these are typical representatives of the peasant bourgeoisie.

Exactly analogous relations are to be found in the starch making industry in the Yuryev Uyezd, Vladimir Gubernia (V. Prugavin, *loc. cit.*, p. 104 and foll.). Here, too, the workshop owners carry on production mainly with the aid of wage-labour (out of 128 workers in 30 workshops, 86 are hired); and here, too, the workshop owners are far above the mass of the peasantry as far as stock-breeding and agriculture are concerned; they use potato pulp as feed for their cattle. Even real capitalist farmers emerge from among the peasants. Mr. Prugavin describes the farm of a peasant who owns a starch works (valued at about 1,500 rubles) employing 12 wage-workers. This peasant grows potatoes on his own farm, which he has enlarged by renting land. The crop rotation is seven-field and includes clover. For the farm work he employs from 7 to 8 workers, hired from spring to autumn ("from end to end"). The pulp is used as cattle feed, and the owner intends to use the waste water for his fields.

Mr. V. Prugavin assures us that this works enjoys "quite exceptional conditions." Of course, in any capitalist society the rural bourgeoisie will always constitute a very small minority of the rural population, and in this sense will, if you like, be an "exception." But this term will not eliminate the fact that in the starch-making area, as in all the other commercial farming areas in Russia, a class of rural entrepreneurs is being formed, who are organising capitalist agriculture.

4) Vegetable-Oil Production

The extraction of oil from linseed, hemp, sunflower and other seeds is also frequently an agricultural industry. One can gauge the development of vegetable-oil production in the post-Reform period from the fact that in 1864 the vegetable-oil output had an estimated value of 1,619,000 rubles, in 1879 of 6,486,000 rubles, and in 1890 of 12,232,000 rubles. In this branch of production, too, a double process of development is to be observed: on the one hand, small peasant (and sometimes also landlord) oil presses producing oil for sale are established in the villages. On the other hand, large steam-driven works develop, which concentrate production and oust the small establishments. Here we are interested solely in the agricultural processing of oil-bearing plants. "The owners of the hempseed oil presses," we read in the *Historico-Statistical Survey* (Vol. II), "belong to the well-to-do members of the peasantry"; they attach particular value to vegetable-oil production because it enables them to obtain excellent feed for their cattle (oil cake).

Mr. Prugavin (*loc. cit.*), noting the "extensive development of the production of linseed oil" in the Yuryev Uyezd, Vladimir Gubernia, states that the peasants derive "no little advantage" from it (pp. 65-66), that crop and stock raising is conducted on a far higher level by peasants who own oil presses than by the bulk of the peasantry and that some of the oil millers also resort to the hire of rural workers (*loc. cit.*, tables, pp. 26-27 and 146-147). The Perm handicraft census for 1894-95 also showed that crop raising is conducted on a much higher level by handicraft oil millers than by the bulk of the peasants (larger areas under crops, far more animals, better harvests, etc.), and that this improvement in cultivation is accompanied by the hiring of rural workers. In the post-Reform period in Voronezh Gubernia, there has been a particular development of the commercial cultivation of sunflower seed, which is crushed for oil in local presses. The area under sunflowers in Russia in the 70s was estimated at about 80,000 dess. (*Historico-Statistical Survey*, I), and in the 80s at about 136,000 dess., of which 2/3 belonged to peasants. "Since then, however, judging by certain data, the area under this plant has considerably increased, in some places by 100 per cent and even more" (*Productive Forces*, I, 37). "In the village of Alexeyevka alone" (Biryuch Uyezd, Voronezh Gubernia), we read in the *Historico-Statistical Survey*, Vol. II, "there are more than 40 oil presses, and Alexeyevka itself, solely owing to sunflowers, has prospered and grown from a wretched little hamlet into a rich town ship, with houses and shops roofed with sheet iron" (p. 41). How this wealth of the peasant bourgeoisie was reflected in the condition of the mass of the peasantry may be seen from the fact that in 1890, in the village of Alexeyevka, out of 2,273 families registered (13,386 persons of both sexes), 1,761 had no draught animals, 1,699 had no implements, 1,480 cultivated no land, and only 33 families did not engage in industries.

Tobacco Growing

In conclusion, let us make some brief observations on the development of tobacco growing. The average crop in Russia for the years 1863-1867 was 1,923,000 poods from 32,161 dess.; for 1872-1878 it was 2,783,000 poods from 46,425 dess.; for the 80s, it was 4 million poods from 50,000 dess. The number of plantations in the same periods was estimated at 75,000, 95,000 and 650,000 respectively, which evidently indicates a very considerable increase in the number of small cultivators drawn into this type of commercial farming. Tobacco growing requires a considerable number of workers. Among the types of agricultural migration note is therefore made of migration to tobacco plantations (particularly to the outer gubernias in the South, where the cultivation of tobacco has recently expanded with exceptional rapidity). Reference has already been made in publications to the fact that the workers on the tobacco plantations lead a very hard life.

In the *Survey of Tobacco Growing in Russia* (Parts II and III, St. Petersburg, 1894, published by order of the Department of Agriculture), there are very detailed and interesting data on tobacco growing as a branch of commercial farming. Mr. V. S. Shcherbachov, describing tobacco growing in Malorossia, gives wonderfully precise information on three uyezds of Poltava Gubernia (Priluki, Lokhvitsa and Romny). This information, gathered by the author and arranged by the Bureau of Statistics, Poltava Gubernia Zemstvo Board, covers 25,089 peasant farms in the three uyezds that grow tobacco; they have 6,844 dessiatines under tobacco and 146,774 dessiatines under cereals. We see an enormous concentration of both the tobacco and the cereal area in the hands of the capitalist farms. Less than one-eighth of the farms (3,000 out of 25,000) hold more than half the area under cereals (74,000 dess. out of 147,000), with an average of nearly 25 dess. per farm. Almost half the area under tobacco (3,200 dess. out of 6,800) belongs to these farms, the average per farm being over 1 dessiatine, whereas for all the other groups the area under tobacco does not exceed one- to two-tenths of a dessiatine per household.

The owner of 7 dessiatines under tobacco must there fore have at least 14 workers; in other words, he must undoubtedly base his farm on wage-labour. Some grades of tobacco require not two but three seasonal workers per dessiatine, and day labourers in addition. In a word, we see quite clearly that the greater the degree to which agriculture becomes commercial, the more highly developed is its capitalist organisation.

The preponderance of small and tiny farms among the tobacco growers (11,997 farms out of 25,089 have *up to one-tenth* of a dessiatine planted) does not in the least refute the fact of the capitalist organisation of this branch of commercial agriculture; for this mass of tiny farms accounts for an insignificant share of the output (11,997, i.e., nearly half the farms, have in all 522 dess. out of 6,844, or less than one-tenth). Nor do "average" figures, to which people so often confine themselves, provide a picture of the real situation (the average per farm is a little over 114 dessiatine under tobacco).

In some uyezds the development of capitalist agriculture and the concentration of production are still more marked. In the Lokhvitsa Uyezd, for example, 229 farms out of 5,957 each have 20 dessiatines and more under cereals. Their owners have 22,799 dess. under cereals out of a total of 44,751, i.e., more than half. Each farmer has about 100 dess. under crops. Of the land under tobacco they have 1,126 dess. out of 2,003 dess. And if the farms are grouped according to area under tobacco, we have in this uyezd 132 farmers out of 5,957 with two and more dessiatines under tobacco. These 132 farmers have 1,441 dess. under tobacco out of 2,003, i.e., 72% and more than *ten* dessiatines under tobacco per farm. At the other extreme of the same Lokhvitsa Uyezd we have 4,360 farms (out of 5,957) having up to one-tenth of a dessiatine each under tobacco, and altogether 133 dessiatines out of 2,003, i.e., 6%.

It goes without saying that the capitalist organisation of *production* is accompanied here by a very considerable development of *merchant's* capital and by all sorts of exploitation outside the sphere of production. The small tobacco growers have no drying sheds, are unable to give their tobacco time to ferment and to sell it (in 3 to 6 weeks) as a finished product. They sell the unfinished product at *half the price* to buyers-up, who very often plant tobacco themselves on rented land. The buyers-up "squeeze the small planters in every way" (p. 31 of cited publication). Commercial agriculture is commercial capitalist production: this relation can be clearly traced (if only one is able to select the proper methods) in this branch of agriculture too.

INDUSTRIAL VEGETABLE AND FRUIT GROWING; SUBURBAN FARMING

With the fall of serfdom, "landlord fruit growing," which had been developed on quite a considerable scale, "suddenly and rapidly fell into decline almost all over Russia." The construction of railways changed the situation, giving a "tremendous impetus" to the development of new, commercial fruit growing, and brought about a "complete change for the better" in this branch of commercial agriculture. On the one hand, the influx of cheap fruit from the South undermined the industry in the centres where it was formerly conducted; and on the other hand, industrial fruit growing developed, for example, in the Kovno, Vilna, Minsk, Grodno, Mogilev and Nizhni-Novgorod gubernias, along with the expansion of the fruit market. Mr. V. Pashkevich points out that an investigation into the condition of fruit farming in 1893-94 revealed a considerable development of it as an industrial branch of agriculture in the previous ten years, an increase in the demand for gardeners, undergardeners, etc. Statistics confirm such views: the amount of fruit carried by the Russian railways is increasing, fruit imports, which increased in the first decade after the Reform, are declining.

It stands to reason that commercial vegetable growing, which provides articles of consumption for incomparably larger masses of the population than fruit growing does, has developed still more rapidly and still more extensively. Industrial vegetable growing becomes widespread, firstly, near the towns ; secondly, near factory and commercial and industrial settlements and also along the railways; and thirdly, in certain villages, scattered throughout Russia and famous for their vegetables. It should be observed that there is a demand for this type of produce not only among the industrial, but also among the agricultural population: let us recall that the budgets of the Voronezh peasants show a per-capita expenditure on vegetables of 47 kopeks, more than half of this expenditure being on *purchased* produce.

To acquaint ourselves with the social and economic relations that arise in this type of commercial agriculture we must turn to the data of local investigations in the particularly developed vegetable-growing areas. Near St. Petersburg, for example, frame and hot-house vegetable growing is widely developed, having been introduced by migratory vegetable growers from Rostov. The number of frames owned by big growers runs into thousands, and by medium growers, into hundreds. "Some of the big vegetable growers supply tens of thousands of poods of pickled cabbage to the army." According to Zemstvo statistics in Petersburg Uyezd 474 households of the local population are engaged in vegetable growing (about 400 rubles income per household) and 230 in fruit growing. Capitalist relations are very extensively developed both in the form of merchant's capital (the industry is "ruthlessly exploited by profiteers") and in the form of hiring workers. Among the immigrant population, for example, there are 115 master vegetable growers (with an income of over 3,000 rubles each) and 711 worker vegetable growers (with an income of 116 rubles each.)

The peasant vegetable growers near Moscow are the same sort of typical members of the rural bourgeoisie. "According to an approximate estimate, over 4 million poods of vegetables and greens reach Moscow's markets every year. Some of the villages do a big trade in pickled vegetables: Nogatino Volost sells nearly a million vedros of pickled cabbage to factories and barracks, and even sends consignments to Kronstadt. . . . Commercial vegetable growing is widespread in all the Moscow uyezds, chiefly in the vicinity of towns and factories." "The cabbage is chopped by hired labourers who come from Volokolamsk Uyezd" (*Historico-Statistical Survey*, I, p. 19).

Exactly similar relations exist in the well-known vegetable-growing district in Rostov Uyezd, Yaroslavl Gubernia, embracing 55 vegetable-growing villages—Porechye, Ugodichi and others. All the land, except pastures and meadows, has long been turned into vegetable fields. The technical processing of vegetables—preserving—is highly developed. Together with the product of the land, the land itself and labour-power are converted into commodities. Despite the "village community," the inequality of land tenure, for example, in the village of Porechye, is very great: in one case a family of 4 has 7 "vegetable plots," in another a family of 3 has 17; this is explained by the fact that no periodical land redistribution takes place here; only private redivisions take place, and the peasants "freely exchange" their "vegetable plots" or "patches" (*Survey of Yaroslavl Gubernia*, 97–98). "A large part of the field-work . . . is done by male and female day labourers, many of whom come to Porechye in the summer season both from neighbouring villages and from neighbouring gubernias" (*ibid.*, 99). It is estimated that in the whole of Yaroslavl Gubernia 10,322 persons (of whom 7,689 are from Rostov) engaged in "agriculture and vegetable growing" are *migratory* workers—i.e., in the majority of cases are wage-workers in the given occupation. The above quoted data on the migration of rural workers to the metropolitan gubernias, Yaroslavl Gubernia, etc., should be brought into connection with the development not only of dairy farming but also of commercial vegetable growing.

Vegetable growing also includes the hot-house cultivation of vegetables, an industry that is rapidly developing among the well-to-do peasants of Moscow and Tver gubernias. In the first-named gubernia the 1880-81 census showed 88 establishments with 3,011 frames; there were 213 workers, of whom 47 (22.6%) were hired; the total output was valued at 54,400 rubles. The average hot-house vegetable grower had to put at least 300 rubles into the "business." Of the 74 peasants for whom house-to-house returns are given, 41 possess purchased land, and as many rent land, there is an average of 2.2 horses per peasant. It is clear from this that the hot-house vegetable industry is only within the reach of members of the peasant bourgeoisie.

In the south of Russia melon growing also comes within the type of commercial agriculture under review. Here are some brief observations about its development in a district described in an interesting article in the *Vestnik Finansov* (1897, No. 16) on "industrial melon growing." This branch of production arose in the village of Bykovo (Tsarev Uyezd, Astrakhan Gubernia) at the end of the 60s and the beginning of the 70s. The melons, which at first went only to the Volga region, were consigned, with the coming of the railways, to the capital cities. In the 80s the output "increased at least tenfold" owing to the enormous profits (150 to 200 rubles per dess.) made by the initiators of the business. Like true petty bourgeois, they did all they could to prevent the number of growers from increasing and were most careful in guarding from their neighbours the "secret" of this new and profitable occupation. Of course, all these heroic efforts of the "muzhik cultivator" to stave off "fatal competition" were in vain, and the industry spread much wider—to Saratov Gubernia and the Don region. The drop in grain prices in the 90s gave a particularly strong impetus to production, compelling "local cultivators to seek a way out of their difficulties in crop rotation systems." The expansion of production considerably increased the demand for hired labour (melon growing requires a considerable amount of labour, so that the cultivation of one dessiatine costs from 30 to 50 rubles), and still more considerably increased the profits of the employers and ground-rent. Near "Log" Station (Gryazi-Tsaritsyn Railway), the area under water-melons in 1884 was 20 dess., in 1890 between 500 and 600 dess., and in 1896 between 1,400 and 1,500 dess., while rent rose from 30 kopeks to between 1.50 and 2 rubles and to between 4 and 14 rubles per dess. for the respective years. The over-rapid expansion of melon planting led at last, in 1896, to overproduction and a crisis, which finally confirmed the capitalist character of this branch of commercial agriculture. Melon prices fell to a point where they did not cover railway charges. The melons were left ungathered in the fields. After tasting tremendous profits the entrepreneurs now learned what losses were like. But the most interesting thing is the means they have chosen for combating the crisis: the means chosen is to win new markets, to effect such a cheapening of produce and of railway tariffs as to transform it from an item of luxury into an item of consumption for the people (and at points of production, into cattle feed). "Industrial melon growing," the entrepreneurs assure us, "is on the road to further development; apart from high railway tariffs there is no obstacle to its further growth. On the contrary, the Tsaritsyn-Tikhoretskaya Railway now under construction . . . opens a new and extensive area for industrial melon growing." Whatever the further destiny of this "industry" may be, at any rate the history of the "melon crisis" is very instructive, constituting a miniature picture, it is true, but a very vivid one, of the capitalist evolution of agriculture.

We still have to say a few words about *suburban farming*. The difference between it and the above-described types of commercial agriculture is that in their case the entire farm is adapted to some one chief market product. In the case of suburban farming, however, the small cultivator trades in bits of everything: he trades in his house by letting it to summer tenants and permanent lodgers, in his yard, in his horse and in all sorts of produce from his fields and farmyard: grain, cattle feed, milk, meat, vegetables, berries, fish, timber, etc.; he trades in his wife's milk (baby-farming near the capitals), he makes money by rendering the most diverse (not always even mentionable) services to visiting townsfolk, etc., etc. The complete transformation by capitalism of the ancient type of patriarchal farmer, the complete subjugation of the latter to the "power of money" is expressed here so vividly that the suburban peasant is usually put in a separate category by the Narodnik who says that he is "no longer a peasant." But the difference between this type and all preceding types is only one of form. The political and economic essence of the all-round transformation effected in the small cultivator by capitalism is everywhere the same. The more rapid the increase in the number of towns, the number of factory, commercial and industrial townships, and the number of railway stations, the more extensive is the area of the transformation of our "village community man" into this type of peasant. We should not forget what was said in his day by Adam Smith—that improved communications tend to convert every village into a suburb. Remote areas cut off from the outside world, already an exception, are with every passing day increasingly becoming as rare as antiquities, and the cultivator is turning with ever-growing rapidity into an industrialist subjected to the general laws of commodity production.

In thus concluding the review of the data on the growth of commercial agriculture, we think it not superfluous to repeat here that our aim has been to examine the main (by no means all) forms of commercial agriculture.

CONCLUSIONS ON THE SIGNIFICANCE OF CAPITALISM IN RUSSIAN AGRICULTURE

In chapters II-IV the problem of capitalism in Russian agriculture has been examined from two angles. First we examined the existing system of social and economic relations in peasant and landlord economy, the system which has taken shape in the post-Reform period. It was seen that the peasantry have been splitting up at enormous speed into a numerically small but economically strong rural bourgeoisie and a rural proletariat. Inseparably connected with this "depeasantising" process is the landowners' transition from the labour-service to the capitalist system of farming. Then we examined this same process from another angle: we took as our starting-point the manner in which agriculture is transformed into commodity production, and examined the social and economic relations characteristic of each of the principal forms of commercial agriculture. It was shown that the very same processes were conspicuous in both peasant and private-landowner farming under a great variety of agricultural conditions.

Let us now examine the conclusions that follow from all the data given above.

1) The main feature of the post-Reform evolution of agriculture is its growing commercial, entrepreneur character. As regards private-landowner farming, this fact is so obvious as to require no special explanation. As regards peasant farming, however, it is not so easily

established, firstly, because the employment of hired labour is not an absolutely essential feature of the small rural bourgeoisie. As we have observed above, this category includes every small commodity-producer who covers his expenditure by independent farming, provided the general system of economy is based on the capitalist contradictions examined in Chapter II. Secondly, the small rural bourgeois (in Russia, as in other capitalist countries) is connected by a number of transitional stages with the small-holding "peasant," and with the rural proletarian who has been allotted a patch of land. This circumstance is one of the reasons for the viability of the theories which do not distinguish the existence of a rural bourgeoisie and a rural proletariat among "the peasantry."

2) From the very nature of agriculture its transformation into commodity production proceeds in a special way, unlike the corresponding process in industry. Manufacturing industry splits up into separate, quite independent branches, each devoted exclusively to the manufacture of one product or one part of a product. The agricultural industry, however, does not split up into quite separate branches, but merely specialises in one market product in one case, and in another market product in another, all the other aspects of agriculture being adapted to this principal (i.e., market) product. That is why the forms of commercial agriculture show immense diversity, varying not only in different areas, but also on different farms. That is why, when examining the question of the growth of commercial agriculture, we must on no account confine ourselves to gross data for agricultural production as a whole.

3) The growth of commercial agriculture creates a home market for capitalism. Firstly, the specialisation of agriculture gives rise to exchange between the various agricultural areas, between the various agricultural undertakings, and between the various agricultural products. Secondly, the further agriculture is drawn into the sphere of commodity circulation the more rapid is the growth of the demand made by the rural population

for those products of manufacturing industry that serve for personal consumption; and thirdly, the more rapid is the growth of the demand for means of production, since neither the small nor the big rural entrepreneur is able, with the old-fashioned "peasant" implements, buildings, etc., etc., to engage in the new, commercial agriculture. Fourthly and lastly, a demand is created for labour-power, since the formation of a small rural bourgeoisie and the change-over by the landowners to capitalist farming presuppose the formation of a body of regular agricultural labourers and day labourers. Only the fact of the growth of commercial agriculture can explain the circumstance that the post-Reform period is characterised by an expansion of the home market for capitalism (development of capitalist agriculture, development of factory industry in general, development of the agricultural engineering industry in particular, development of the so-called peasant "agricultural industries," i.e., work for hire, etc.).

4) Capitalism enormously extends and intensifies among the agricultural population the contradictions without which this mode of production cannot exist. Notwithstanding this, however, agricultural capitalism in Russia, in its historical significance, is a big progressive force. Firstly, capitalism has transformed the cultivator from a "lord of the manor," on the one hand, and a patriarchal, dependent peasant, on the other, into the same sort of *industrialist* that every other proprietor is in present-day society. Before capitalism appeared, agriculture in Russia was the business of the gentry, a lord's hobby for some, and a duty, an obligation for others; consequently, it could not be conducted except according to age-old routine, necessarily involving the complete isolation of the cultivator from all that went on in the world beyond the confines of his village. The labour-service system—that living survival of old times in present-day economy—strikingly confirms this characterisation. Capitalism for the first time broke with the system of social estates in land tenure by converting the land into a commodity. The farmer's

product was put on sale and began to be subject to social reckoning—first in the local, then in the national, and finally in the international market, and in this way the former isolation of the uncouth farmer from the rest of the world was completely broken down. The farmer was compelled willy nilly, on pain of ruin, to take account of the sum-total of social relations both in his own country and in other countries, now linked together by the world market. Even the labour-service system, which formerly guaranteed Oblomov an assured income without any risk on his part, without any expenditure of capital, without any changes in the age-old routine of production, now proved incapable of saving him from the competition of the American farmer. That is why one can fully apply to post-Reform Russia what was said half a century ago about Western Europe—that agricultural capitalism hag been "the motive force which has drawn the idyll into the movement of history."

Secondly, agricultural capitalism has for the first time undermined the age-old stagnation of our agriculture; it has given a tremendous impetus to the transformation of its technique, and to the development of the productive forces of social labour. A few decades of "destructive work" by capitalism have done more in this respect than entire centuries of preceding history. The monotony of routine natural economy has been replaced by a diversity of forms of commercial agriculture; primitive agricultural implements have begun to yield place to improved implements and machines; the immobility of the old-fashioned farming systems has been undermined by new methods of agriculture. The course of all these changes is linked inseparably with the above-mentioned phenomenon of the specialisation of agriculture. By its very nature, capitalism in agriculture (as in industry) cannot develop evenly: in one place (in one country, in one area, on one farm) it pushes forward one aspect of agriculture, in another place another aspect, etc. In one case it transforms the technique of some, and in other cases of other agricultural operations, divorcing them from patriarchal peasant economy or from the patriarchal labour-service. Since the whole of this process is guided by market requirements that are capricious and not always known to the producer, capitalist agriculture, in each separate instance (often in each separate area, sometimes even in each separate country), becomes more one sided and lopsided than that which preceded it, but, taken as a whole, becomes immeasurably more many-sided and rational than patriarchal agriculture. The emergence of separate types of commercial agriculture renders possible and inevitable capitalist crises in agriculture and cases of capitalist overproduction, but these crises (like all capitalist crises) give a still more powerful impetus to the development of world production and of the socialisation of labour.

Thirdly, capitalism has for the first time created in Russia large-scale agricultural production based on the employment of machines and the extensive co-operation of workers. Before capitalism appeared, the production of agricultural produce was always carried on in an unchanging, wretchedly small way—both when the peasant worked for himself and when he worked for the landlord—and no "community character" of land tenure was capable of destroying this tremendously scattered production. Inseparably linked with this scattered production was the scattered nature of the farmers themselves. Tied to their allotment, to their tiny "village community," they were completely fenced off even from the peasants of the neighbouring village community by the difference in the categories to which they belonged (former landowners' peasants, former state peasants, etc.), by differences in the size of their holdings—by differences in the terms on which their emancipation took place (which terms were sometimes determined simply by the individual attributes of the landlords and by their whims). Capitalism for the first time broke down these purely medieval barriers—and it was a very good thing that it did. Now the differences between the various grades of peasants, between the various categories based on the size of allotment holdings, are far less important than the economic differences within each grade, each category and each village community. Capitalism destroys local seclusion and insularity, and replaces the minute medieval divisions among cultivators by a major division, embracing the whole nation, that divides them into classes occupying different positions in the general system of capitalist economy. The mass of cultivators were formerly tied to their place of residence by the very conditions of production, whereas the creation of diverse forms and diverse areas of commercial and capitalist agriculture could not but cause the movement of enormous masses of the population throughout the country; and unless the population is mobile (as we have said above) there can be no question of developing its understanding and initiative.

Fourthly, and lastly, agricultural capitalism in Russia for the first time cut at the root of labour-service and the personal dependency of the farmer. This system of labour-service has held undivided sway in our agriculture from the days of *Russkaya Pravda* down to the present day cultivation of the fields of private landowners with the peasants' implements; the wretchedness and uncouthness of the farmer, degraded by his labour being "semi-free" if not feudal, in character, are inevitable concomitants of this system; if the civil rights of the cultivator had not been impaired (by, for example, his belonging to the lowest social estate; corporal punishment; assignment to public works; attachment to allotment, etc.) the labour-service system would have been impossible. That is why agricultural capitalism in Russia has performed a great historical service in replacing labour-service by hired labour. Summing up what has been said above on the progressive historical role of Russian agricultural capitalism, it may be said that it is socialising agricultural production. Indeed, the fact that agriculture has been transformed from the privileged occupation of the top estate or the duty of the bottom estate into an ordinary commercial and industrial occupation; that the product of the cultivator's labour has become subject to social reckoning on the market; that routine, uniform agriculture is being converted into technically transformed and diverse forms of commercial farming; that the local seclusion and scattered nature of the small farmers is breaking down; that the diverse forms of bondage and personal dependence are being replaced by impersonal transactions in the purchase and sale of labour-power, these are all links in a single process, which is socialising agricultural labour and is increasingly intensifying the contradiction between the anarchy of market fluctuations, between the individual character of the separate agricultural enterprises and the collective character of large-scale capitalist agriculture.

Thus (we repeat once more), in emphasising the progressive historical role of capitalism in Russian agriculture we do not in the least forget either the historically transient character of this economic regime or the profound social contradictions inherent in it. On the contrary, we have shown above that it is precisely the Narodniks who, capable only of bewailing the "destructive work" of capitalism, give an extremely superficial appraisal of these contradictions, glossing over the differentiation of the peasantry, ignoring the capitalist character of the employment of machinery in our agriculture, and covering up with such expressions as "agricultural industries" and "employments" the emergence of a class of agricultural wage-workers.

2

Capitalism and Agriculture

Capitalism is an economic system in which wealth, and the means of producing wealth, are privately owned and controlled rather than commonly, publicly, or state-owned and controlled. Through capitalism, the land, labor, and capital are owned, operated, and traded by private individuals either singly or jointly, and investments, distribution, income, production, pricing and supply of goods, commodities and services are determined by voluntary private decision in a market economy. A distinguishing feature of capitalism is that each person owns his or her own labor and therefore is allowed to sell the use of it to employers. In a "capitalist state", private rights and property relations are protected by the rule of law of a limited regulatory framework. In the modern capitalist state, legislative action is confined to defining and enforcing the basic rules of the market, though the state may provide some public goods and infrastructure.

Some consider laissez-faire to be "pure capitalism." Laissez-faire (French, "leave to do (by itself)"), signifies minimizing or eliminating state interference in economic affairs and the competitive process, allowing the free play of supply and demand. Laissez-faire capitalism has never existed in practice. Because all large economies today have a mixture of private and public ownership and control, some feel that the term "mixed economies" more precisely describes most contemporary economies. In the "capitalist mixed economy", the state intervenes in market activity and provides many services.

During the last century, capitalism has often been contrasted with centrally planned economies. The central axiom of capitalism is that the best allocation of resources is achieved through consumers having free choice, and producers responding accordingly to meet collective consumer demand. This contrasts with planned economies in which the state directs what shall be produced.

A consequence is the belief that privatization of previously state-provided services will tend to achieve a more efficient delivery thereof. Further implications are usually in favor of free trade, and abolition of subsidies. Although individuals and groups must act rationally in any society for their own good, the consequences of both rational and irrational actions are said to be more readily apparent in a capitalist society.

Capitalistic economic practices incrementally became institutionalized in England between the 16th and 19th centuries, although some features of capitalist organization existed in the ancient world, and early aspects of merchant capitalism flourished during the Late Middle Ages. Capitalism has been dominant in the Western world since the end of feudalism. From Britain, it gradually spread throughout Europe, across political and cultural frontiers. In the 19th and 20th centuries, capitalism provided the main, but not exclusive, means of industrialization throughout much of the world.

ETYMOLOGY

The etymology of the word *capital* has roots in the trade and ownership of animals. The Latin root of *capital* is *capitalis*, from the proto-Indo-European *kaput*, which means "head", this being how wealth was measured—the number of heads in a person's livestock. The terms chattel and cattle itself also derive from this same origin.

The lexical connections between animal trade and economics can also be seen in the names of many currencies and words about money: fee (faihu), rupee (rupya), buck (a deerskin), pecuniary (pecu), stock (livestock), and peso (pecu or pashu) all derive from animal-trade origins.

Although Adam Smith is often described as the "father of capitalist thinking," he never used the term "capitalism". He described his own preferred economic system as "the system of natural liberty." However, Smith defined "capital" as stock, and "profit" as the just expectation to keep the revenue from improvements to that stock. Smith also made capital improvement the central goal of the economic and political system.

According to the *Oxford English Dictionary*, *capitalism* was first used by novelist William Makepeace Thackeray in 1854, by which he meant by having ownership of capital.

According to the *OED*, Carl Adolph Douai, a German-American socialist and abolitionist, used the term *private capitalism* in 1863.

Karl Marx and Friedrich Engels referred the *capitalistic system* (*kapitalistischen System*) to the capitalist mode of production (*kapitalistische Produktionsform*) in *Das Kapital* (1867). The word "capitalism" only appears twice in *Das Kapital*, namely in tome II, p.124 (German edition) and in *Theories about Surplus Value*, tome II, p.493 (German edition). However, the late Engels made more frequent use of the term "capitalism". Marx's notion of the capitalist mode of production is characterised as a system of primarily private ownership of the means of production in a mainly market economy, with a legal framework on commerce and a physical infrastructure provided by the state.

An 1877 work entitled *Better Times*, and an unknown author in 1884 of the *Pall Mall Gazette*, also used the term *capitalism*.

However, the first use of *capitalism* to describe the production system was by the German economist Werner Sombart, in his 1902 book *The Jews and Modern Capitalism* (*Die Juden und das Wirtschaftsleben*). Sombart's close friend and colleague, Max Weber, also used *capitalism* in his 1904 book *The Protestant Ethic and the Spirit of Capitalism* (*Die protestantische Ethik und der Geist des Kapitalismus*).

Perspectives

The concept of capitalism has evolved over time, with later thinkers often building on the analysis of earlier thinkers. Moreover, the component concepts used in defining capitalism — such as private ownership, markets and investment — have evolved along with changes in theory, in law, and in practice. This is a concept that is often compared with laborism.

CLASSICAL POLITICAL ECONOMY

The classical school economic thought emerged in Britain in the late 18th century. The classical political economists Adam Smith, David Ricardo, Jean-Baptiste Say, and John Stuart Mill published analyses of the production, distribution and exchange of goods in a market that have since formed the basis of study for most contemporary economists.

In France, 'Physiocrats' like François Quesnay promoted free trade based on a conception that wealth originated from land. Quesnay's *Tableau Économique* (1759), described the economy analytically and laid the foundation of the Physiocrats economic theory, followed by Anne Robert Jacques Turgot who opposed tariffs and customs duties and advocated free trade. Richard Cantillon defined long-run equilibrium as the balance of flows of income, and argued that the supply and demand mechanism around land influenced short-term prices.

Adam Smith's attack on mercantilism and his reasoning for "the system of natural liberty" in *The Wealth of Nations* (1776) are usually taken as the beginning of classical political economy. Smith devised a set of concepts that remain strongly associated with capitalism today, particularly his theory of the "invisible hand" of the market, through which the pursuit of individual self-interest unintentionally produces a collective good for society. It was necessary for Smith to be so forceful in his argument in favor of free markets because he had to overcome the popular mercantilist sentiment of the time period. He criticized monopolies, tariffs, duties, and other state enforced restrictions of his time and believed that the market is the most fair and efficient arbitrator of resources. This view was shared by David Ricardo, second most important of the classical political economists and one of the most influential economists of modern times. In *The Principles of Political Economy and Taxation* (1817), he developed the law of comparative advantage, which explains why it is profitable for two parties to trade, even if one of the trading partners is more efficient in every type of economic production. This principle supports the economic case for free trade. Ricardo was a supporter of Say's Law and held the view that full employment is the normal equilibrium for a competitive economy. He also argued that inflation is closely related to changes in quantity of money and credit and was a proponent of the law of diminishing returns, which states that each additional unit of input yields less and less additional output.

The values of classical political economy are strongly associated with the classical liberal doctrine of minimal government intervention in the economy, though it does not necessarily oppose the state's provision of a few basic public goods.. Classical liberal thought has generally assumed a clear division between the economy and other realms of social activity, such as the state.]

While economic liberalism favors markets unfettered by the government, it maintains that the state has a legitimate role in providing public goods. For instance, Adam Smith argued that the state has a role in providing roads, canals, schools and bridges that cannot be efficiently implemented by private entities. However, he preferred that these goods should be paid proportionally to their consumption (e.g. putting a toll). In addition, he advocated retaliatory tariffs to bring about free trade, and copyrights and patents to encourage innovation.

WEBERIAN POLITICAL SOCIOLOGY

In some social sciences, the understanding of the defining characteristics of capitalism has been strongly influenced by 19th century German social theorist Max Weber. Weber considered market exchange, rather than production, as the defining feature of capitalism; capitalist enterprises, in contrast to their counterparts in prior modes of economic activity, was their rationalization of production, directed toward maximizing efficiency and productivity. According to Weber, workers in pre-capitalist economic institutions understood work in terms of a personal relationship between master and journeyman in a guild, or between lord and peasant in a manor.

In his book *The Protestant Ethic and the Spirit of Capitalism* (1904-1905), Weber sought to trace how capitalism transformed these traditional modes of economic activity. For Weber, the 'spirit of capitalism' began with the Puritan understanding of one's 'calling' in life and their laboring for God rather than for men. This is pictured in Proverbs 22:29, "Seest thou a man diligent in his calling? He shall stand before kings" and in Colossians 3:23, "Whatever you do, do your work heartily, as for the Lord rather than for men." In the *Protestant Ethic*, Weber further stated that "moneymaking – provided it is done legally – is, within the modern economic order, the result and the expression of diligence in one's calling..." Thus in Weber's opinion, it was with a devotion to God in the workplace and seeking assurance of salvation described as the Protestant work ethic that the Puritans helped form the basis to the modern economic order.

This 'spirit' was gradually codified by law; rendering wage-laborers legally 'free' to sell work; encouraging the development of technology aimed at the organization of production on the basis of rational principles; and clarifying the apparent separation of the public and private lives of workers, especially between the home and the workplace. Therefore, unlike Marx, Weber did not see capitalism as primarily the consequence of changes in the means of production.

Capitalism, for Weber, was the most advanced economic system ever developed over the course of human history. Weber associated capitalism with the advance of the business corporation, public credit, and the further advance of bureaucracy of the modern world. Although Weber defended capitalism against its socialist critics of the period, he saw its rationalizing tendencies as a possible threat to traditional cultural values and institutions, and a possible 'iron cage' constraining human freedom. This is further seen in his criticism of "specialists without spirit, hedonists without a heart" that were developing, in his opinion, with the fading of the original Puritan 'spirit' associated with capitalism.

GERMAN HISTORICAL SCHOOL AND AUSTRIAN SCHOOL

From the perspective of the German Historical School, capitalism is primarily identified in terms of the organization of production for markets. Although this perspective shares similar theoretical roots with that of Weber, its emphasis on markets and money lends it different focus. For followers of the German Historical School, the key shift from traditional modes of economic activity to capitalism involved the shift from medieval restrictions on credit and money to the modern monetary economy combined with an emphasis on the profit motive. In the late 19th century, the German Historical School of economics diverged, with the emerging Austrian School of economics, led at the time by Carl Menger. Later generations of followers of the Austrian School continued to be influential in Western economic thought through much of the 20th century. The Austrian economist Joseph Schumpeter, a forerunner of the Austrian School of economics, emphasized the "creative destruction" of capitalism — the fact that market economies undergo constant change. At any moment of time, posits Schumpeter, there are rising industries and declining industries. Schumpeter, and many contemporary economists influenced by his work, argue that resources should flow from the declining to the expanding industries for an economy to grow, but they recognized that sometimes resources are slow to withdraw from the declining industries because of various forms of institutional resistance to change.

The Austrian economists Ludwig von Mises and Friedrich Hayek were among the leading defenders of market capitalism against 20th century proponents of socialist planned economies. Mises and Hayek argued that only market capitalism could manage a complex, modern economy. Since a modern economy produces such a large array of distinct goods and services, and consists of such a large array of consumers and enterprises, asserted Mises and Hayek, the information problems facing any other form of economic organization other than market capitalism would exceed its capacity to handle information. Thinkers within Supply-side economics built on the work of the Austrian School, and particularly emphasize Say's Law: "supply creates its own demand." Capitalism, to this school, is defined by lack of state restraint on the decisions of producers.

Austrian economists claim that Marx failed to make the distinction between *capitalism* and *mercantilism*. They argue that Marx conflated the imperialistic, colonialistic, protectionist and interventionist doctrines of mercantilism with capitalism.

Austrian economics has been a major influence on the ideology of libertarianism, which considers laissez-faire capitalism to be the ideal economic system. Murray Rothbard, who founded the Center for Libertarian Studies and the Journal of Libertarian Studies, is referred to as the father of Libertarianism in the United States. He was associated with the 1982 creation of the Ludwig von Mises Institute and later was its academic vice president. In 1987 he started the scholarly "Review of Austrian Economics," now called the Quarterly Journal of Austrian Economics. Rothbard coined the term "Anarcho-capitalism".

KEYNESIAN ECONOMICS

In his 1937 *The General Theory of Employment, Interest, and Money*, the British economist John Maynard Keynes argued that capitalism suffered a basic problem in its ability to recover from periods of slowdowns in investment. Keynes argued that a capitalist economy could remain in an indefinite equilibrium despite high unemployment. Essentially rejecting Say's law, he argued that some people may have a liquidity preference which would see them rather hold money than buy new goods or services, which therefore raised the prospect that the Great Depression would not end without what he termed in the *General Theory* "a somewhat comprehensive socialization of investment."

Keynesian economics challenged the notion that laissez-faire capitalist economics could operate well on their own, without state intervention used to promote aggregate demand, fighting high unemployment and deflation of the sort seen during the 1930s. He and his followers recommended "pump-priming" the economy to avoid recession: cutting taxes, increasing government borrowing, and spending during an economic down-turn. This was to be accompanied by trying to control wages nationally partly through the use of inflation to cut real wages and to deter people from holding money. John Maynard Keynes tried to provide solutions to many of Marx's problems without completely abandoning the classical understanding of capitalism. His work attempted to show that regulation can be effective, and that economic stabilizers can reign in the aggressive expansions and recessions that Marx disliked. These changes sought to create more stability in the business cycle, and reduce the abuses of laborers. Keynesian economists argue that Keynesian policies were one of the primary reasons capitalism was able to recover following the Great Depression.The premises of Keynes's work have, however, since been challenged by neoclassical and supply-side economics and the Austrian School. Another challenge to Keynesian thinking came from his colleague Piero Sraffa, and subsequently from the Neo-Ricardian school that followed Sraffa. In Sraffa's highly technical analysis, capitalism is defined by an entire system of social relations among both producers and consumers, but with a primary emphasis on the demands of production. According to Sraffa, the tendency of capital to seek its highest rate of profit causes a dynamic instability in social and economic relations.

Neoclassical Economics and the Chicago School

Today, most academic research on capitalism in the English-speaking world draws on neoclassical economic thought. It favors extensive market coordination and relatively neutral patterns of governmental market regulation aimed at maintaining property rights, rather than privileging particular social actors; deregulated labor markets; corporate governance dominated by financial owners of firms; and financial systems depending chiefly on capital market-based financing rather than state financing.

Milton Friedman effectively took many of the basic principles set forth by Adam Smith and the classical economists and modernized them, in a way. One example of this is his article in the September 1970 issue of The New York Times Magazine, where he claims that the social responsibility of business is "to use its resources and engage in activities designed to increase its profits...(through) open and free competition without deception or fraud." This is tantamount to Smith's argument that self interest in turn benefits the whole of society. Work like this helped lay the foundations for the coming remarketization of capitalism and the supply-side economics of Ronald Reagan and Margaret Thatcher.

The Chicago School of economics is best known for its free market advocacy and monetarist ideas. According to Milton Friedman and monetarists, market economies are inherently stable if left to themselves and depressions result only from government intervention. Friedman, for example, argued that the Great Depression was result of a contraction of the money supply, controlled by the Federal Reserve, and not by the lack of investment as Keynes had argued. Ben Bernanke, current

Chairman of the Federal Reserve, is among the economists today generally accepting Friedman's analysis of the causes of the Great Depression.

Neoclassical economists, today the majority of economists, consider value to be subjective, varying from person to person and for the same person at different times, and thus reject the labor theory of value. Marginalism is the theory that economic value results from marginal utility and marginal cost (the marginal concepts). These economists see capitalists as earning profits by forgoing current consumption, by taking risks, and by organizing production.

History

Mercantilism

The period between the 16th and 18th centuries is commonly described as mercantilism. This period was associated with geographic discoveries by merchant overseas traders, especially from England and the Low Countries; the European colonization of the Americas; and the rapid growth in overseas trade. Mercantilism was a system of trade for profit, although commodities were still largely produced by non-capitalist production methods. While some scholars see mercantilism as the earliest stage of modern capitalism, others argue that modern capitalism did not emerge until later. For example, noting the pre-capitalist features of mercantilism, Karl Polanyi argued that capitalism did not emerge until the establishment of free trade in Britain in the 1830s.

The earliest forms of mercantilism date back to the Roman Empire. When the Roman Empire expanded, the mercantilist economy expanded throughout Europe. After the collapse of the Roman Empire, most of the European economy became controlled by local feudal powers, and mercantilism collapsed there. However, mercantilism persisted in Arabia. Due to its proximity to neighboring countries, the Arabs established trade routes to Egypt, Persia, and Byzantium. As Islam spread in the 7th century, mercantilism spread rapidly to Spain, Portugal, Northern Africa, and Asia. Mercantilism finally revived in Europe in the 14th century, as mercantilism spread from Spain and Portugal. Feudalism began to lay some of the foundations necessary for the development of mercantilism, a precursor to capitalism. Feudalism took place mostly in Europe and lasted from the medieval period up through the 16th century. Feudal manors were almost entirely self sufficient, and therefore limited the role of the market. This stifled the growth of capitalism. However, the relatively sudden emergence of new technologies and discoveries, particularly in the industries of agriculture and exploration, revitalized the growth of capitalism. The most important development at the end of Feudalism was the emergence of "the dichotomy between wage earners and capitalist merchants".

Among the major tenets of mercantilist theory was bullionism, a doctrine stressing the importance of accumulating precious metals. Mercantilists argued that a state should export more goods than it imported so that foreigners would have to pay the difference in precious metals. Mercantilists asserted that only raw materials that could not be extracted at home should be imported; and promoted government subsides, such as the granting of monopolies and protective tariffs, were necessary to encourage home production of manufactured goods. European merchants, backed by state controls, subsidies, and monopolies, made most of their profits from the buying and selling of goods. In the words of Francis Bacon, the purpose of mercantilism was "the opening and well-balancing of trade; the cherishing of manufacturers; the banishing of idleness; the repressing of waste and excess by sumptuary laws; the improvement and husbanding of the soil; the regulation of prices..." Similar practices of economic regimentation had begun earlier in the medieval towns. However, under mercantilism, given the contemporaneous rise of the absolutism, the state superseded the local guilds as the regulator of the economy. During that time the guilds essentially functioned like cartels that monopolized the quantity of craftsmen to earn above-market wages.

Commercialism

At the period of the 18th century, the commercial stage of capitalism transcended from the previous domination of capitalism by merchants. Commercialism, or commercial capitalism, originated from the start of the British and Dutch East India Company. These companies were characterized by its monopoly on trade, granted by the letters patents. Recognized as chartered joint-stock companies by the state, these companies enjoyed a large sum of power, ranging from lawmaking, military, and treaty-making privileges. Characterized by its colonial and expansionary powers by states, powerful nation-states sought to accumulate precious metals, and military conflicts arose. During this era, merchants, who had traded under the previous stage of mercantilism, invested capital in the East India Companies and other colonies, seeking a return on investment.

Industrialism

By the late 18th century, mercantilism was in crisis: mercantile activity could not produce sufficient wealth to pay for the military expenditures of the states that protected, and depended on, commerce. This crisis intensified with the Industrial Revolution. Although mercantilist policies endured in European countries with weak industrial bases, such as Prussia and Russia, into the 19th century, rapidly industrializing countries began questioning the value of mercantilist policies by the late 18th century. This is most evident in Great Britain, the home of the Industrial Revolution, where a new group of economic theorists, led by David Hume and Adam Smith, in the mid 18th century, challenged fundamental mercantilist doctrines as the belief that the amount of the world's wealth remained constant and that a state could only increase its wealth at the expense of another state.

At the same time that philosophers and politicians were debating the merits of mercantilism, the mid-18th century gave rise to an alternative set of economic relations and practices: industrial (bourgeois) capitalism. Most scholars agree that the emergence of capitalism was made possible by earlier economic developments in England. According to Marxists, it was made possible by the exploitation of wage-labor on a large scale, which English landowners first experimented with after the crisis of the 14th century. According to World Systems Theorists like Immanuel Wallerstein, it was made possible by the accumulation of vast amounts of capital under the merchant phase of capitalism.

During the resulting Industrial Revolution, the industrialist replaced the merchant as a dominant actor in the capitalist system and effected the decline of the traditional handicraft skills of artisans, guilds, and journeymen. Also during this period, capitalism marked the transformation of relations between the British landowning gentry and peasants, giving rise to the production of cash crops for the market rather than for subsistence on a feudal manor. The surplus generated by the rise of commercial agriculture encouraged increased mechanization of agriculture and the rise of the bourgeoisie.

Marx dated industrial capitalism from the last third of the 18th century, marked the development of the factory system of manufacturing, characterized by a complex division of labor between and within work process and the routinization of work tasks; and finally established the global domination of the capitalist mode of production. In the midst of this newly developing concept of division of labor came exploitation of labor on a much larger scale than was ever seen before.

Britain also abandoned its protectionist policy, as embraced by mercantilism. In the 19th century, Richard Cobden and John Bright, who based their beliefs on the Manchester School, initiated a movement to lower tariffs. In the 1840s, Britain adopted a less protectionist policy, with the repeal of the Corn Laws and the Navigation Acts. Britain reduced tariffs and quotas, in line with Adam Smith and David Ricardo's advocacy for free trade. As noting the various pre-capitalist features of mercantilism, Karl Polanyi argued that capitalism did not emerge until the establishment of free trade in Britain in the 1830s. Other sources indicate that mercantilism fell after the repeal of the Navigation Acts in 1849, and libertarians argue that the current system is still mercantilist.

However, due to companies legislation, British capitalism was not exclusively laissez-faire. The British state created charters, creating immunites for the coporations under the Limited Liability Act 1855 and the Joint Stock Companies Act 1856. The British East India Company and controls in major industries during that time were also important examples of economic regulations.

Most of the early proponents of the liberal theory of economics in the United States subscribed to the American School. This school of thought was inspired by the ideas of

Alexander Hamilton, who proposed the creation of the First National Bank and the Second National Bank and increased tariffs (e.g. tariff of 1828) to favor northern industrial interests. Following Hamilton's death, the more abiding protectionist influence in the antebellum period came from Henry Clay and his *American System*.

In the mid-19th century, the United States followed the Whig tradition of economic liberalism, which included increased state control, regulation and macroeconomic development of infrastructure. Public works such as the provision and regulation transportation such as railroads took effect. The Pacific Railway Acts provided the development of the First Transcontinental Railroad. In order to help pay for its war effort in the American Civil War, the United States government imposed its first personal income tax, on August 5, 1861, as part of the Revenue Act of 1861 (3% of all incomes over US $800; rescinded in 1872).

Following the American Civil War, the movement towards a mixed economy accelerated with even more protectionism and government regulation. In the 1880s and 1890s, significant tariff increases were enacted. Moreover, with the enactment of the Interstate Commerce Act of 1887, the Sherman Anti-trust Act, the federal government began to assume an increasing role in regulating and directing the country's economy.

Monopolism

In the late 19th century, the control and direction of large areas of industry came into the hands of financiers. This period has been defined as state capitalism, state monopoly capitalism, or corporate capitalism, characterized by the subordination of processes of production to the accumulation of profits in a financial system. Major characteristics of capitalism in this period included the establishment of large industrial cartels or monopolies; the ownership and management of industry by financiers divorced from the production process; and the development of a complex system of banking, an equity market, and corporate holdings of capital through stock ownership. Increasingly, large industries and land became the subject of profit and loss by financial speculators.

From about the American Civil War to the early 20th century, capitalism has also been increasingly influenced by large, monopolistic corporations. The oil, telecommunication, railroad, shipping, banking and financial industries are characterized by its monopolistic domination. Inside these corporations, a division of labor separates shareholders, owners, managers, and actual laborers. Although the concept of monopoly capitalism originated among Marxist theorists, non-Marxist economic historians have also commented on the rise of monopolies and trusts in the period. Stromberg, Joseph R. *(2001).* "The Role of State Monopoly Capitalism in the American Empire". *Journal of Libertarian Studies 15 (3): 74–75.* http://mises.org/journals/jls/15_3/15_3_3.pdf. ">

By the last quarter of the 19th century, the emergence of large industrial trusts had provoked legislation in the U.S. to reduce the monopolistic tendencies of the period. Gradually, during this Progressive Era, the U.S. federal government played a larger and larger role in passing antitrust laws and regulation of industrial standards for key industries of special public concern.

However, contemporary, non-bourgeois economic historians believe these new laws were in fact designed to aid large corporations at the expense of smaller competitors. By the end of the 19th century, economic depressions and boom and bust business cycles had become a recurring problem, although such problems were most likely caused by government intervention, not failures in free markets (Rand 1967, Friedman 1962, Bernstein 2005).

In particular, the Long Depression of the 1870s and 1880s and the Great Depression of the 1930s affected almost the entire capitalist world, and generated discussion about capitalism's long-term survival prospects. In the early 20th century, a succession of U.S. Presidents, beginning with Warren Harding's "Return to Normalcy," the state decreased taxation rates, with the Revenue Act of 1924 and 1926.

This allowed for the prosperity of "The Roaring Twenties," but later was said to be largely responsible for the Great Depression. During the 1930s, Marxist commentators often posited the possibility of capitalism's decline or demise, often in alleged contrast to the ability of the Soviet Union to avoid suffering the effects of the global depression.

Keynesianism and Neoliberalism

In the period following the global depression of the 1930s, the state played an increasingly prominent role in the capitalistic system throughout much of the world. In 1929, for example, total U.S. government expenditures (federal, state, and local) amounted to less than one-tenth of GNP; from the 1970s they amounted to around one-third (EB). Similar increases were seen in all bourgeois economies, some of which, such as France, have reached even higher ratios of government expenditures to GNP than the United States. These economies have since been widely described as "mixed economies."

During the postwar boom, a broad array of new analytical tools in the social sciences were developed to explain the social and economic trends of the period, including the concepts of post-industrial society and the welfare state. The phase of capitalism from the beginning of the postwar period through the 1970s has sometimes been described as "state capitalism", especially by Marxian thinkers. This era was greatly influenced by Keynesian economic stabilization policies.

The long postwar boom ended in the late 1960s and early 1970s, and the situation was worsened by the rise of stagflation. Exceptionally high inflation combined with slow output growth, rising unemployment, and eventually recession caused loss of credibility of Keynesian welfare-statist mode of regulation.

Under the influence of Friedrich Hayek and Milton Friedman, Western states embraced policy prescriptions inspired by the laissez-faire capitalism and classical liberalism. In particular, monetarism, a theoretical alternative to Keynesianism that is more compatible with laissez-faire, gained increasing prominence in the capitalist world, especially under the leadership of Ronald Reagan in the U.S. and Margaret Thatcher in the UK in the 1980s.

Finally, the general public's interest shifted from the collectivist concerns of Keynes's managed capitalism to a focus on individual freedom and choice, called "remarketized capitalism." In the eyes of many economic and political commentators, collapse of the Soviet Union supposedly brought further evidence of superiority of market capitalism over communism.

Globalization

Although overseas trade has been associated with the development of capitalism for over five hundred years, some thinkers argue that a number of trends associated with globalization have acted to increase the mobility of people and capital since the last quarter of the 20th century, combining to circumscribe the room to maneuver of states in choosing non-capitalist models of development. Today, these trends have bolstered the argument that capitalism should now be viewed as a truly world system. However, other thinkers argue that globalization, even in its quantitative degree, is no greater now than during earlier periods of capitalist trade. The roots of globalized capitalism can be traced back to the imperialism of the early 20th century. Imperialistic policies promoted the spread of capitalistic principles, and the doors of trade stayed open in foreign countries even after imperialism had come to an end.

After the abandonment of the Bretton Woods system and the strict state control of foreign exchange rates, the total value of transactions in foreign exchange was estimated to be at least twenty times greater than that of all foreign movements of goods and services (EB). The internationalization of finance, which some see as beyond the reach of state control, combined with the growing ease with which large corporations have been able to relocate their operations to low-wage states, has posed the question of the 'eclipse' of state sovereignty, arising from the growing 'globalization' of capital.

While scientists generally agree about the size of global income inequality, there is a general disagreement about the recent direction of change of it. However, it is growing within particular nations such as China. The book *The Improving State of the World* argues that economic growth since the Industrial Revolution has been very strong and that factors such as adequate nutrition, life expectancy, infant mortality, literacy, prevalence of child labor, education, and available free time have improved greatly.

The biggest reason for the increasingly global capitalist economy is the telecommunications revolution that has taken place over the last twenty years. Fax machines, cell phones, and the internet have made it possible for work to be done and transactions to take place from almost anywhere in the world.

In 2008, state intervention in global capital markets by the American and other governments was seen by many as signaling a crisis for free-market capitalism. Serious turmoil in the banking system and financial markets due in part to the subprime mortgage crisis reached a critical stage during September 2008, characterized by severely contracted liquidity in the global credit markets and going-concern threats to investment banks and other institutions.

POLITICAL ADVOCACY

Support

Many theorists and policymakers in predominantly capitalist nations have emphasized capitalism's ability to promote economic growth, as measured by Gross Domestic Product (GDP), capacity utilization or standard of living. This argument was central, for example, to Adam Smith's advocacy of letting a free market control production and price, and allocate resources. Many theorists have noted that this increase in global GDP over time coincides with the emergence of the modern world capitalist system. While the measurements are not identical, proponents argue that increasing GDP (per capita) is empirically shown to bring about improved standards of living, such as better availability of food, housing, clothing, and health care. The decrease in the number of hours worked per week and the decreased participation of children and the elderly in the workforce have been attributed to capitalism. Proponents also believe that a capitalist economy offers far more opportunities for individuals to raise their income through new professions or business ventures than do other economic forms. To their thinking, this potential is much greater than in either traditional feudal or tribal societies or in socialist societies.

Milton Friedman has argued that the economic freedom of competitive capitalism is a requisite of political freedom. Friedman argued that centralized control of economic activity is always accompanied by political repression. In his view, transactions in a market economy are voluntary, and the wide diversity that voluntary activity permits is a fundamental threat to repressive political leaders and greatly diminish power to coerce. Friedman's view was also shared by Friedrich Hayek and John Maynard Keynes, both of whom believed that capitalism is vital for freedom to survive and thrive.

Austrian School economists have argued that capitalism can organize itself into a complex system without an external guidance or planning mechanism. Friedrich Hayek coined the term "catallaxy" to describe what he considered the phenomenon of self-organization underpinning capitalism. From this perspective, in process of self-organization, the profit motive has an important role. From transactions between buyers and sellers price systems emerge, and prices serve as a signal as to the urgent and unfilled wants of people. The promise of profits gives entrepreneurs incentive to use their knowledge and resources to satisfy those wants. Thus the activities of millions of people, each seeking his own interest, are coordinated.

This decentralized system of coordination is viewed by some supporters of capitalism as one of its greatest strengths. They argue that it permits many solutions to be tried, and that real-world competition generally finds a good solution to emerging challenges. In contrast, they argue, central planning often selects inappropriate solutions as a result of faulty forecasting. However, in all existing modern economies, the state conducts some degree of centralized economic planning (using such tools as allowing the country's central bank to set base interest rates), ostensibly as an attempt to improve efficiency, attenuate cyclical volatility, and further particular social goals. Proponents who follow the Austrian School argue that even this limited control creates inefficiencies because we cannot predict the long-term activity of the economy. Milton Friedman, for example, has argued that the Great Depression was caused by the erroneous policy of the Federal Reserve.

Ayn Rand was a prominent philosophical supporter of laissez-faire capitalism; her novel Atlas Shrugged was one of the most influential publications ever written on the subject of business and continues to be a best-seller. The first person to endow capitalism with a new code of morality (Rational Selfishness), she did not justify capitalism on the grounds of pure "practicality" (that it is the best wealth-creating system), or the supernatural (that God or religion supports capitalism), or because it benefits the most people, but maintained that it is the only morally valid socio-political system because it allows people to be free to act in their rational self-interest. These thinkers have had a substantial influence on the Libertarian Party. The Libertarian Party strongly advocates the elimination of most, if not all, state involvement in the marketplace. The Republican Liberty Caucus is the libertarian branch of the Republican Party.

Criticism

Prominent leftist critics have included socialists (like Karl Marx, Friedrich Engels, Vladimir Lenin, Mao Zedong, Leon Trotsky, Antonio Gramsci, Rosa Luxemburg, Slavoj Zizek, Che Guevara, Fidel Castro) and anarchists (including Benjamin Tucker, Lysander Spooner, Pierre-Joseph Proudhon, Mikhail Bakunin, Peter Kropotkin, Emma Goldman, Murray Bookchin, Rudolf Rocker, Noam Chomsky, and many others).

Movements like the Luddites, Narodniks, Shakers, Utopian Socialists and others have opposed capitalism for various reasons. Marxism advocated a revolutionary overthrow of capitalism that would lead eventually to communism. Marxism also influenced social democratic and labour parties, which seek change through existing democratic channels instead of revolution, and believe that capitalism should be heavily regulated rather than abolished. Many aspects of capitalism have come under attack from the relatively recent anti-globalization movement.

Some religions criticize or outright oppose specific elements of capitalism. Some traditions of Judaism, Christianity, and Islam forbid lending money at interest, although methods of Islamic banking have been developed. Christianity has been a source of both praise and criticism for capitalism, particularly its materialist aspects.

The first socialists drew many of their principles from Christian values "bourgeois" values of profiteering, greed, selfishness, and hoarding. Christian critics of capitalism may not oppose capitalism entirely, but support a mixed economy in order to ensure adequate labor standards and relations, as well as economic justice. In addition, there are many prominent Protestant denominations (particularly in the United States) who have reconciled with — or are ardently in favor of —

capitalism, particularly in opposition to secular socialism. However, in the U.S. and around the world there are many Protestant Christian traditions which are critical of, or even oppose, capitalism. Another critic is the Indian philosopher P.R. Sarkar, founder of the Ananda Marga movement, who developed the Social Cycle Theory as a solution to the problems of capitalism called the Progressive Utilization Theory (PROUT).

Some problems said to be associated with capitalism include: unfair and inefficient distribution of wealth and power; a tendency toward market monopoly or oligopoly (and government by oligarchy); imperialism and various forms of economic and cultural exploitation; and phenomena such as social alienation, inequality, unemployment, and economic instability.

Critics have maintained that there is an inherent tendency towards oligolopolistic structures when laissez-faire is combined with capitalist private property.

Because of this tendency either laissez-faire, or private property, or both, have drawn fire from critics who believe an essential aspect of economic freedom is the extension of the freedom to have meaningful decision-making control over productive resources to everyone.

Economist Branko Horvat asserts, "it is now well known that capitalist development leads to the concentration of capital, employment and power. It is somewhat less known that it leads to the almost complete destruction of economic freedom." SMU Economics Professor and New York Times #1 best-selling author, Ravi Batra, has long maintained that excessive income and wealth inequalities are a fundamental cause of financial crisis and economic depression in the capitalist economy.

Near the start of the 20th century, Vladimir Lenin argued that that state use of military power to defend capitalist interests abroad was an inevitable corollary of monopoly capitalism. This concept of political economy concerning the relationship between economic and political power among and within states includes critics of capitalism who assign to it responsibility for not only economic exploitation, but imperialist, colonialist and counter-revolutionary wars, repressions of workers and trade unionists, genocides, massacres, and so on.

Some environmentalists argue that capitalism requires continual economic growth, and will inevitably deplete the finite natural resources of the earth, and other broadly utilized resources. Such thinkers, including Murray Bookchin, have argued that capitalist production externalizes environmental costs to all of society, and is unable to adequately mitigate its impact upon ecosystems and the biosphere at large. Supporters maintain, however, that it would be imprudent for capitalist societies to deplete resources to such an extent.

Some labor historians and scholars, such as Immanuel Wallerstein, Tom Brass and, latterly Marcel van der Linden, have also argued that unfree labor — the use of a labor force comprised of slaves, indentured servants, criminal convicts, political prisoners, and/or other coerced persons — is compatible with capitalist relations.

DEMOCRACY, THE STATE, AND LEGAL FRAMEWORKS

The relationship between the state, its formal mechanisms, and capitalist societies has been debated in many fields of social and political theory, with active discussion since the 19th century. Hernando de Soto is a contemporary economist who has argued that an important characteristic of capitalism is the functioning state protection of property rights in a formal property system where ownership and transactions are clearly recorded. According to de Soto, this is the process by which physical assets are transformed into capital, which in turn may be used in many more ways and much more efficiently in the market economy. A number of Marxian economists have argued that the Enclosure Acts in England, and similar legislation elsewhere, were an integral part of capitalist primitive accumulation and that specific legal frameworks of private land ownership have been integral to the development of capitalism.

New institutional economics, a field pioneered by Douglass North, stresses the need of capitalism for a legal framework to function optimally, and focuses on the relationship between the historical development of capitalism and the creation and maintenance of political and economic institutions. In new institutional economics and other fields focusing on public policy, economists seek to judge when and whether governmental intervention (such as taxes, welfare, and government regulation) can result in potential gains in efficiency. According to Gregory Mankiw, a New Keynesian economist, governmental intervention can improve on market outcomes under conditions of "market failure," or situations in which the market on its own does not allocate resources efficiently.

The idea of market failure is that markets fail to realize all potential gains from trade. This means that markets fail to deliver perfect economic results. Critics of market failure theory, like Ronald Coase, Harold Demsetz, and James M. Buchanan argue that government programs and policies also fall short of absolute perfection. Market failures are often small, and government failures are sometimes large. It is therefore the case that imperfect markets are often better than imperfect governmental alternatives. While all nations currently have some kind of market regulations, the desirable degree of regulation is disputed.

The relationship between democracy and capitalism is a contentious area in theory and popular political movements. The extension of universal adult male suffrage in 19th century Britain occurred along with the development of industrial capitalism, and democracy became widespread at the same time as capitalism, leading many theorists to posit a causal relationship between them, or that each affects the other.

However, in the 20th century, according to some authors, capitalism also accompanied a variety of political formations quite distinct from liberal democracies, including fascist regimes, monarchies, and single-party states, while it has been observed that many democratic societies such as the Bolivarian Republic of Venezuela and Anarchist Catalonia have been expressly anti-capitalist.

While some thinkers argue that capitalist development more-or-less inevitably eventually leads to the emergence of democracy, others dispute this claim. Research on the democratic peace theory further indicates that capitalist democracies rarely make war with one another and have little internal violence. However critics of the democratic peace theory note that democratic capitalist states may fight infrequently or never with other democratic capitalist states because of Political similarity or political stability rather than because they are democratic (or capitalist).

Some commentators argue that though economic growth under capitalism has led to democratization in the past, it may not do so in the future. Under this line of thinking, authoritarian regimes have been able to manage economic growth without making concessions to greater political freedom.

In response to criticism of the system, some proponents of capitalism have argued that its advantages are supported by empirical research. For example, advocates of different Index of Economic Freedom point to a statistical correlation between nations with more economic freedom (as defined by the Indices) and higher scores on variables such as income and life expectancy, including the poor in these nations.

3

The World Capitalist Economy

INTRODUCTION

The decades since the war have been marked by extremely far-reaching changes of world historical significance in all spheres of human affairs. Never before has society experienced such substantial transformations in its socio–economic and political structure in so short a time. Taking them globally, two dialectically linked trends attract attention first of all.

One is the forming and rapid growth of a world system of socialism through the revolutionary replacement of capitalist and pre-capitalist social relations by socialist ones in a number of countries. The other is the steady deepening of capitalism's contradictions within the contracting limits of the non-socialist world. Both these processes provide objective prerequisites for further development of the general crisis of capitalism, an inherent, decisive factor of which is the crisis of the world capitalist economy.

Since World War! II there have been very substantial changes in this economy compared with earlier stages that have affected very important aspects of the economic affairs and international relations of all the countries of the nonsocialist world that comprise it, though by no means to the same extent. Most of them are occurring directly or indirectly under the influence of comparatively new trends in the economic competition of the two world social systems. These trends in turn inevitably reflect the long-term consequences of the break-up of the colonial system, the ever increasing, extremely contradictory influences of the scientific and technical revolution on the productive forces and structure of social production of the various groups of countries, qualitatively new phenomena as regards internationalisation of their social production and the international division of 6 labour, the mounting struggle of the emancipated countries against neocolonial exploitation, for real equality in international economic relations.

The economic and social instability of the world capitalist system became unusually intense at the end of the period we study, especially as a result of the marked sharpening in the 70s of the energy, raw material, inflation, cyclic, international financial, and other major problems of world capitalist economy. These crisis processes, which had a logical continuation and development in the years following, developed in a complicated, eventful situation of intense struggle between two lines in world politics, viz., on the one hand, that of curbing the arms race, consolidating peace and detente, and defending the sovereign rights and freedoms of nations pursued by world socialism and the whole peaceloving mankind, and, on the other hand, the imperialist line of undermining detente and stepping up the arms race, and policy of threats and interference in others' affairs and suppression of the national liberation struggle.

Capitalism, as was stressed at the 26th Congress of the CPSU, has not, of course, congealed in the present conditions of its general crisis; from the beginning of the 70s through to the early 80s alone, it has experienced three economic recessions in spite of capitalist governments' counter-crisis measures and all their efforts to prevent new slumps in production. Inflation has attained an unprecedented scale in these countries. As the Central Committee's report to the Congress noted:

> It is more than obvious that slate regulation of the capitalist economy is ineffective. The measures that bourgeois governments take against inflation foster stagnation of production and growth of unemployment; what they do to contain the critical drop in production lends still greater momentum to inflation.

The more that facts are accumulated revealing critical trends in, and specific features of, the growth of the productive forces in the economies of the main groups of countries of world capitalism, the more pressing is it becoming to systematise, generalise, and interpret them in the aggregate and in their inter-relationship. A problem of fundamental importance is arising, viz., whether a complex approach to the tackling of such a broad task is possible at all. For the aggregate of facts on the economic activity of all countries and of all the industries of world capitalism (in their dynamics and interconditioning, and their repeatedly intersecting national and international connections) embraces a truly immense range of matters, each of which plays a far from identical role in the development both of the national states, and the industries and spheres of their economies, taken separately, and of the world economy as a whole.

The answer to this question was first given by Marxism; and it consisted not so much in the discovery of some allembracing method for summarising a vast number of facts and trends as in its bringing out the theoretical and methodological premises of a systems approach to analysis of the inner logic and patterns of the capitalist system's functioning at the various concrete stages of its 'self-movement'. Lenin stressed, when speaking of the changes constantly occurring in the capitalist world economy, that the sumtotal of these changes in all their ramifications in the capitalist world economy could not bo grasped even by seventy Marxes. The most important thing is that the *laws* of these changes have been discovered, that the *objective* logic of these changes and of their historical development has in its chief and basic features been disclosed. It is its understanding of the objective laws of social development that has enabled Marxist-Leninist economic science to tackle elaboration of the principles of a scientific methodology for analysing capitalism as a world economic system.

The characteristics of the most important stages of the long forming and evolution of this system taken together with the genesis of the capitalist mode of production, the world market, and the aims and methods of colonial exploitation of some countries by others, constitute the main substance of Part I of this book, in which special attention is paid to defining the object of study and the internal contradictions and long-term trends of its development that shape the basic causes of the rise and inevitable deepening of the crisis of the world capitalist economy today. The data adduced will provide the necessary theoretical foundation for bringing ont the themes of the succeeding chapters.

The general results and decisive trends of the development of the postwar world capitalist economy are examined in Part II, above all the ecanomic factors and social consequences of the long-term shifts in the structure and dynamics of both the social production and the consumption of all the countries of this economy taken together. Then follows a detailed comparison of two groups of countries (viz., capitalist and developing) in order to bring out the specific character of their postwar development, including the impact of the scientific and technical revolution and break-up of the colonial system on their economies. The trends in the growth of population and manpower have been analysed from the same standpoint, and also the productivity of social labour in the main geographical regions of the non-socialist world. All this has enabled us to attempt a more circumstantial and detailed estimate of the changes that have come to light in the regional structure of the distribution of the leading sectors of social production in both the industrial centres and the agrarian and primary producer periphery of capitalism. The problem of the cyclic movement of the postwar world capitalist economy has a place of fundamental importance in the book. Analysis of it helps clarify many features of the regular intensification in recent decades of the uneven economic growth and structural instability of modern capitalism's international business relations.

The main lines of the intertwining development of the industrial and agrarian and raw material base of the postwar world capitalist economy are analysed in Part III. Study of the changes in the manufacturing industries in capitalist and developing countries that are most significant as regards their socio-economic consequences occupy the foreground. Then the long-term shifts in the dynamics, volume and structure of the production of raw materials and in the movement of prices of industrial items, farm produce and raw materials on the world capitalist market are examined. On that basis a number of objectively operating trends are brought out, in which the essential features of the present stage of the crisis of the imperialist system of international division of labour are being more and more clearly manifested within the framework of the world economic relations of recent years. It would not be legitimate to appraise all the trends studied in isolation from the decisive patterns of the world 9 revolutionary process of modern times. As Leonid Brezhnev stressed at the 25th Congress of the Communist Party of the Soviet Union:

This is an epoch of radical social change. Socialism's positions arc expanding and growing stronger. The victories of the national liberation movement are opening up now horizons for countries that have won independence. The class struggle of the working people against monopoly oppression, against the exploiting order, is gaining in intensity. The scale of the revolutionary-democratic, anti-imperialist movement is steadily growing. Taken as a whole, this signifies development of the world revolutionary process.

This book does not claim, of course, to be an exhaustive analysis of the problems posed; many of them certainly need further study, tidying up, and concretisation. Because of the inadequacy or scrappiness of reliable statistics and general economic data, separate propositions of the themes are reviewed in outline or are advanced tentatively. When the conclusions are based on published work or well-known data the exposition has been compressed to a resume, and references are given to the sources containing the detailed argumentation.

The Russian original appeared in 1978. In the present English edition the author has tried to continue analysis of the problems and trends studied, employing the latest data available to him. On that basis certain amendments have been made, and the conclusions and estimates of the probable outlook for the development of the world capitalist economy have been elaborated and broadened.

METHODOLOGICAL AND HISTORICAL PREMISES

The Subject-matter of the Study

The Basic Principles of the Analysis

The economic content of historical progress is extremely broad and many-sided, but at all stages of mankind's development it has ultimately been determined by the objective needs of increasing the productive forces of society. The growth of production and deepening of the social division of labour stimulated by these needs provide the necessary material preconditions for passing from lower to higher socio-economic formations, and for the succession of one historical epoch by another. As Karl Marx wrote in the preface to the first German edition of *Capital*, from his standpoint the evolution of the economic formation of society is viewed as a process of natural history.

That fundamental theoretical proposition, which made it possible for the first time to put study of the inner logic of social development on a sound scientific basis, was the outcome of the investigation and generalisation of a truly gigantic amount of factual material. Horizons new in principle were thereby opened up for the social sciences, political economy included. At the same time this inference by no means implied denial of the significance of subjective factors for society's progressive movement. It followed that the sole real possibility of a scientific understanding of the role and place of subjective factors in mankind's material and spiritual affairs arises only from study of the objective patterns of the movement of social production. As Marx indicated:

My standpoint ... can less than any other make thn individual responsible for relations whose creature lie socially remains, however much he may subjectively raise himself above them.

The most important of the tasks facing Marx in his investigation of the then dominant historical formation was study of the laws of capitalist production operating and materialised, as he put it, with iron necessity. By characterising a law as an inner and necessary connection between phenomena, he demonstrated in detail that only the disclosure of that kind of inter-relation, manifested in the form of a tendency, leads to understanding of the objective processes of social development. That conception in turn has to serve as the point of departure for man's mastery of the laws of this development, and for his conscious, purposeful activity. Social activity, moreover, as the founders of MarxismLeninism demonstrated in detail, is decisively mediated by the interaction and struggle of the classes that arise and develop on the basis of the material conditions already created at any given period.

The Marxian principles of appraising the mainsprings of the historical process have a universal character. In other words, they are the initial socio-economic premises of a theoretical analysis of the patterns of social development both of countries taken separately and of mankind as a whole. The fundamental nature of these principles is coming out now with extraordinary clarity, when the world socialist system, and the scientific and technical revolution developing within the context of the productive forces of society as a whole, are having an ever increasing effect on growth of mankind's production capability, and on extension and consolidation of both the economic and the scientific and technical interconnections of the world economy of the present transitional period.

It was the discovery and investigation of the patterns of development of social production that first placed study of the productive forces, and of the socio-economic structure of society, on a genuine scientific basis, and made it possible to bring out the inner causes of this development as a process of natural history.

Only the reduction of social relations to production relations and of the latter to the level of the productive forces (Lenin stressed), provided a firm basis for the conception that the development of formations of society is a process of natural history. And it goes without saying thai without such a view there can be no social science.

Clarification of the ways capitalism took shape as a world system is an integral component of Marxist-Leninist political economy. Marx, who attached the greatest importance to this system's international economic interconnections, proposed to complete his fundamental work on the anatomy of capitalist society with a comprehensive analysis of matters embracing the international relations of capitalist production and the world market, including world economic crises. In his scheme for subsequent work on *Capital*, he wrote:

The disposition of material has evidently to he made in such a way that [section] one comprises general abstract definitions, which therefore appertain in some measure to all social formations, but in the sense set forth earlier. Two, the categories which constitute the internal structure of bourgeois society and on which the principal classes are based. Capital, wage-labour, landed property, and their relations to one another. Town and country. The three large social classes; exchange between them. Circulation. The (private) credit system. Three, the State as the epitome of bourgeois society. Analysis of its relations to itself. The 'unproductive' classes. Taxes. National debt. Public-credit. Population. Colonies. Emigration. Four, international conditions of production. International division of labour. International exchange. Export and import. Rate of exchange. Five, world market and crises.

This task, titanic in its conception, was not completed, especially its concluding points. Nevertheless, in the volumes of *Capital* already written, and in a number of other works, Marx succeeded in comprehending and defining the initial theoretical premises of the investigation of many decisive aspects of the problems behind the forming of capitalism's world economic connections. Otherwise it would have been impossible to define the place of this formation in sufficient detail in the series of successive exploiter formations. Study of the patterns of these relations makes it possible to analyse the causes of their rise, development, and flourishing more deeply, and also to bring out with sufficient scientific reliability the irreconcilable contradictions of the capitalist mode of production that predetermine its historically transitional character.

The rapidity of the forming of a world system of capitalism in the last decades of the nineteenth century made it more and more necessary to continue the systems analysis begun by Marx of the interaction of the indissolubly linked domestic and international relations of production. Such an analysis was also essential because of the rise of qualitatively new phenomena in the capitalist system through the change in it from free competition to the domination of monopolies, and the growing over of capitalism into a new, imperialist stage.

At the time, too, political division of the world was completed. A period was established in world politics of the omnipotence of the monopoly capital of the colonial powers, between whom a fierce struggle developed for world hegemony and to redivide the already shared out colonies, semi–colonies, and spheres of influence. As a result even the most economically backward countries, in which various forms of precapitalist social relations still predominated, were drawn into the maelstrom of the international interconnections of capitalist production.

By the beginning of the twentieth century capital had completed the formation of a world economic system. Investigation of it became a most pressing task of the social sciences, above all of political economy. There arose a need for further development of Marxist economic theory in the light of the new facts; and it is not fortuitous that an all–embracing description of this economic system in its monopoly stage in fact became the cornerstone of Lenin's theory of imperialism.

In the foreword to his *Imperialism, the Highest Stage of Capitalism*, written already after the October Socialist Revolution in Russia, Lenin specially stressed that the main purpose of the book was, and remains, to present, on the basis of the"summarised returns of irrefutable bourgeois statistics, and the admissions of bourgeois scholars of all countries, a *composite picture* of the world capitalist system in its international relationships at the beginning of the twentieth century—on the eve of the first world imperialist war.

There was then no broader and more complicated object of study in the economic sciences, and could not be. Its study is comparable, at the very least, to analysis of an object consisting of a host of sub-systems each of which in turn includes a countless number of 'cells' that are in constant change and motion.

Any of the sub-systems of the world capitalist economy, or of its component'cells', can and undoubtedly should bo examined separately, taking into account in the main only the inner stimuli of its motion, abstracting from its interaction with the whole. This approach is justified and makes it possible to bring out both the specific features of the given part's development and certain of the concrete manifestations of the general patterns. As regards the capitalist world economy, however, and in general any other macro-economic system, it does not make it possible to characterise the whole object of study as a unity with any, in any way adequate, fullness, and it can lead, moreover, to a garbled representation of its patterns even when one and the same phenomenon is observed in various countries or sectors of the economy.

In studying the capitalist world economy the investigator is faced, in all its acuteness, with the need to single out from a boundless sea of facts and intersecting processes those real tendencies that in their aggregate and interconnection actually determine the whole picture of its development in a given historical situation, and that enable him to judge the main directions of the movement of the world economy, and not partial or superficial events and phenomena, or those specific to the economic situation of separate countries, the diversity of which cannot be taken into consideration by any theory. Lenin drew special attention in *Imperialism, the Highest Stage of Capitalism* to the point that, in order to characterise the objective economic position of the whole system of capitalism and its international relations, one must not take examples or isolated data (in view of the extreme complexity of the phenomena of social life it is always possible to select any number of examples or separate data lo prove any proposition), but *all* the data on the *basis* of economic life in *all* the belligerent countries and the *whole* world.

The key to theoretical study of the aggregate of the processes that characterise the main trends of development of the capitalist economic system was Marx's discovery of the paramount importance of production. It is in concrete study of the patterns of the movement of social production, and of its class character and inseparable unity with the realm of market relations and of distribution and consumption, that the living soul of Marxist-Leninist political economy consists.

Any economic theory remains sterile or still-born if it is not based on detailed analysis of the facts and arises in isolation from them. It will not be able to explain the real tendencies and, consequently, to discover the objective course of social development. At the same time the concreteness of the study of any economic system, however wide, and even more of such a many-sided one as the capitalist world economy, certainly presupposes clarification and systematisation of the most essential processes in their dynamics, and determination of their hierarchical interdependence. In other words, coucreteness presupposes in this case the need to employ the method of scientific abstraction, i.e. abstraction which, in Lenin's words, reflects nature 'more deeply, truly and *completely*'. In this connection definition of initial premises of the methodology of a study of the comprehensive development of real economic trends becomes very important.

It is typical of capitalist economics, in contrast to Marxist, to try and put markets and market relations at the centre of its analysis and not production and the class relations immanently inherent in it. The former are looked upon as determinant or self-sufficing, allegedly developing of themselves and exerting a dominant influence on the production process. The idea of the primariness of the sphere of circulation, with very substantial differences in the evaluation of the main moments of the functioning of economic systems of any scale, has dominated capitalist economics throughout its history. True, its original theoretical postulates are now being criticised more and more often by many leading Western economists, people like Samuelson, Galbraith, Myrdal, Erhard, Leontief, Balassa, and Tinbergen.

Still, this idea has continued in fact to permeate capitalist literature in one form or another since World War 11, whether it deals witJi capitalism's internal or world economic problems. In their detailed work on the growth of the international economy, Kenwood and Lougheed declare the central sphere of their study to be that of circulation, since, as they write:

The exchange ol goods and services is the means through which independent economic units enter into economic relations with one another and become part of local or national economic community. An exchange passes beyond a country's boundaries, national economic systems become parts of broader regional, continental or world economy.

Prof. P. T. Ellsworth stresses, in his textbook, which was widely used in the United States and Great Britain in the 50s and 60s, that the very term 'international economies' indicates 'that it concerns itself with the economic relations between nations'. Such a posing of the matter is quite typical of present-day Western economic literature.

When investigating capitalism's world economic problems, capitalist economists do not, of course, ignore theoretical analysis of the international aspects of the production process, and study many of them very thoroughly, particularly the dynamics of its productive forces, comparations of the level of economic development of various countries and regions, problems of the international division of labour, capital exports, production and technological competition, integrational processes within the various international economic groupings, migration of labour, and so on.

All these processes, however, arc willy-nilly analysed from ideological positions that pursue definite class interests of the capitalists. The world economy is studied mainly from the standpoint of the determining effect of the trends of world supply and demand on production, and of opportunities for effective employment of the market mechanism of state monopoly methods of regulating international economic relations. Objectively the main job of this approach is 18 to conceal the antagonistic essence of the international relations of production under capitalism. Capitalist economists' endeavour to survey the world capitalist economy out of the context of its organic link with the patterns of development of the corresponding exploiter relations of production, and to represent it simply as a mechanically formed aggregate of national markets within the context of the international exchange of goods and services, also serves the same purpose. On this basis capitalist political economy has time and again exerted great efforts to revive the gimcrack idea whose essence is that modern capitalism is still ahle to generate adequate conditions for future 'harmonious' development of the world economy and in the long run to eradicate its deepest inherent defects and contradictions.

Marxist-Leninist economics considers this matter from fundamentally different theoretical positions. It substantiates in detail the proposition that the world capitalist economy, like a commodity economy of any scale (whether local, regional, or national), includes such inseparably interconnected and jointly subordinated elements as production, markets, and a system of socio-economic relations arising from a corresponding division of labour between economic units and producers. Although market factors of exchange and distribution figure primarily on the surface in capitalism's international economic interconnections, it is not they, however, but the social relations in the production process, that play the basic role when the antagonistic, class nature and decisive trends of development of capitalism's world economic system are investigated.

The capitalist mode of production (Marx wrote) is ... a historical means of developing the material forces of production and creating an appropriate world market and is, at the same time, a continual conflict uetwcen this its historical task and its own corresponding relations of social production.

This conclusion has acquired special methodological significance for analysing the deep-seated crisis processes in the modern world capitalist market. It provides an opportunity for a scientific systematising and comprehensive investigation of the objectively operating patterns and irreconcilable contradictions of capitalism's international market relations, which stem from the socio-economic nature of this mode of production.

The Capitalist Mode of Production and the World Market

The continuous interaction of the development of production and the world market had already come out in the early stages of the rise of the capitalist social formation, and the determinant place in this interaction, moreover, was taken by capitalist industry. As a result of his thorough investigation of extensive historical data, Marx came to the conclusion that when in the 16th, and partially still in the 17th century the sudden expansion of commerce and emergence of a new world market overwhelmingly contributed to the fall of the old mode of production and the rise of capitalist production, this was accomplished conversely on the basis of the already existing capitalist mode of production. The world market itself forms the basis for this mode of production. On the other hand, the immanent necessity of this mode of production to produce on an ever-enlarged scale tends to extend the world market continually, so that it is not commerce in this case which revolutionises industry, but industry which constantly revolutionises commerce.

Marxist political economy consequently looks upon the rise and development of the world capitalist market as a constituent of a broader process, i.e. the gradual conversion of capitalism into a universal socio-economic system. This methodological approach has enabled it to disclose the unequal character of the division of labour between countries that is inherent in this system, and the exploiter nature of commodity exchange and of capitalist international economic relations in general on the world market, based as they are on private property in the instruments and means of production.

The struggle for foreign markets has always been an urgent economic need for capitalism, since a striving to extend its sphere of influence without limit is inherently characteristic of it, in contrast to previously dominant modes of production. Because of the operation of objective economic laws, capitalist enterprise inevitably outgrows the boundaries of the community, region, or country, the economic exclusiveness and isolation of countries being broken down by commodity circulation itself.

As a result, capitalist production becomes inconceivable without foreign trade and world economic relations at a certain stage of its development, which in turn reflects growth of the social division of labour on an international scale. Marx stressed that capitalist production rests on the *value* or the transformation of the labour embodied in the product into social labour. But (his is only [possible] on the basis of foreign trade and of the world market. This is at, once the precondition and the result of capitalist production.

He expressed the same idea even more succinctly in the formula 'capitalist production does not exist at all without foreign commerce'.

All that in no way signifies, of course, that any capitalist enterprise indispensably works for an external market and sends what it produces beyond the local or national economy. At the same time it constantly experiences the effect of the already existing system of international business relations and is drawn into their orbit, if not directly, then indirectly. The capitalist, even the one who is wholly oriented on the home market, is forced to compare his own costs of production with world prices as well as with the market prices in his country. An influence of the same order has to be allowed for when an all-round appraisal is being made of the position of labour power in capitalist countries, including the spheres where its value depends wholly or predominantly on domestic factors of production.

Capitalism's international economic interconnections arose later, historically, than its domestic ones, the latter being original and basic. Their level and trend depended to a decisive degree on how far each nation developed its own 21 productive forces, division of labour, and internal dealings. That, however, does not eliminate the undoubted fact that extension of both the home and foreign market under capitalism represents different aspects of a single process; the boundaries between them are very mobile and arbitrary. As the world market develops the problem of the limits of the home market of each country taken separately proves to lie very tightly bound up with that of extension of its external economic relations. Hence one of the essential principles of Marxian researcli into the development of capitalism in scope and depth is that, when analysing international relations of production (including the capitalist class's colonial policy), it is not important where the boundary between the home and foreign market is drawn; what is important is that capitalism cannot exist and develop without constantly expanding the sphere of its domination, without colonising new countries and drawing old non-capitalist countries into the whirlpool of world economy.

The rise of capitalism's international exploiter relations is inextricably linked with the forming of a system of world economy appropriate to them.

The watershed on this point between Marxian and capitalist science is clear and distinct. Although the creation of this system was completed on the whole under imperialism, it by no means follows that its present-day history can be separated from the preceding history of the development of the capitalist mode of production and that imperialist exploitation of some countries by others begins only with capitalism's transition from free enterprise to the domination of monopolies. Many of the features of the international division of labour typical of the monopoly stage had already arisen and begun to take on a stable, long-term character during the transition of capitalist society to the machine stage of production. They were not brought about by chance, transitional factors but wore caused by deep-seated changes in both the home and external conditions of capitalist reproduction.

The analysis of international economic inter-relationships made in their day by the founders of Marxist political economy indicated that it was large-scale machine industry that prepared the ground for the subsequent formation of a world economy. Machine production, as Marx and Engels pithily put it, produced world history for the first time, insofar as it made all civilised nations and every individual member of them dependent for the satisfaction of their wants on the whole world, thus destroying the former natural exclusiveness of separate nations.

It is through the international division of labour on the world market that the tendency to bring countries together economically, constitutionally inherent in capitalism, is manifested. The kernel of this pattern, however, discovered by Marx, is that this tendency, while progressive in itself, led at the same time, because of capitalist methods and the forms of economic association of the different countries, to consolidation of the division of the world into industrially developed countries, and economically backward agrarian and primary commodity producing countries exploited by them. The antagonism between these two main groups of countries, which are opposite poles of the world capitalist economy, has become one of its main contradictions.

This contradiction has become specially acute in the stage of imperialism, and a very important factor in subsequent origin and development of the crisis of the colonial structure of the capitalist economy, a crisis that is irreversible for the monopoly capitalist class. The break-up of the colonial system after World War II, and the marked deepening of the internal contradictions of world capitalism, which are undermining the historically formed foundations of its system of unequal international economic relations, have put on today's agenda, with all urgency, the issue of eliminating the exploiter system of division of labour that completely dominated the world economy in the past.

It would be wrong to treat this task (which corresponds to the vital interests of all nations without exception) as the need to liquidate the progress already made in the internationalisation of social production. The anti-imperialist struggle for genuine national independence, just like the transition of countries to a higher socio-economic system, does not eliminate the objective need to develop the social (including the international) division of labour, and leads to a change in and perfecting of the forms in which this essentially progressive process is manifested. It is self-evident, Marx wrote to Ludvig Kugelman on 11 July 1868, that this necessity of the distribution of social labour in definite proportions cannot possibly be done away with by a *particular jiirin* of social production but can only change the mode of *its appearance*.

The underlying principles of a Marxist-Leninist analysis of the ways and causes of capital's internationalisation of social production have not lost their theoretical value over the years. The topicality of many of them, moreover, is growing, especially in today's situation of the coexistence and opposition of two world social systems and of the natural extension of economic and technical co-operation on the basis of mutual advantage and equal international division of labour.

In this situation capitalist ideologists are trying in every way to prove the 'erroneousness', in particular, or at least 'obsoleteness' of the Marxist-Leninist political economy's analysis. Above all they endeavour to refute its conclusions about the irreversibility of tendencies toward a steady deepening of capitalism's internal contradictions. But the whole long history of capitalist economics, and of its mistakes, its constant wavering between one conception and another, its fallacies and failures, shows that theories that consider the international market relations of capitalism to be legitimate and that are abstracted from a socio-economic analysis of the production sphere, at best contain superficial analogues of the real trends of development. In the main they bring out the consequences but not the causal connections of this development.

At the same time it would clearly he an oversimplification of things to consider that the theoretical findings of capitalist political economy in the sphere of the world economy's market relations are in general isolated from real analysis of the reproduction processes. In a numher of cases they more or less adequately reflect important features of the backlash of these relations in the production sphere. And this effect determines many of the essential features of the capitalist economy at the various stages of its development.

In the last instance (Engols wrote to Conrad Schmidt on 27 October 1890) production is the decisive factor. But as soon as trade in products becomes independent of production proper, it has a movement of its own, which, although by and large governed by that of production, nevertheless in particulars and within this general dependence again follows laws of its own inherent in the nature of this new' factor; this movement has phases of its own and in its turn reacts on the movement of production.

All this calls for close attention to research both into the ways and determinant tendencies in the genesis of the world market and into its place in the process of extended capitalist reproduction. Not only is such research of educational value historically; it is also very necessary for an analysis of current changes in the structure of the world capitalist economy. A systems analysis of its inner and external interconnections in today's international situation cannot be made sufficiently full without detailed allowance for tiie general patterns and long-term trends of development of these interconnections.

Lenin more than once stressed the need to study broad socio-economic phenomena in historical retrospect, since the most important thing if one is to approach this question scientifically is not to forget the underlying historical connection, to examine every question from the standpoint of how the given phenomenon arose in history and what were the principal stages in its development, and, from the standpoint of its development to examine what it has become today.

There were relatively broad trade and financial relations between separate countries and peoples, of course, in the socio-economic formations that preceded capitalism, but it was capitalism that first formed a world market which played a progressive, revolutionary role in the subsequent increase of mankind's productive forces and growth of its social production. Marx, attributing paramount importance to that, wrote that 'the specific task of bourgeois society is the establishment of a world market'. It was with the formation of a world market that the level of development of capitalism began to be marked by the degree of growth both of domestic and of international market relations.

There was a qualitatively new leap in the shaping of these relations associated with the vast increase in trade and colonial expansion of the young European capitalist class. This expansion became a decisive factor in the primitive accumulation of capital. The world market thus created important material preconditions for victory of the capitalist modeof production over the feudal.

At that time the merchant capital of European powers undoubtedly prevailed in international economic relations, but its dominance on the world market began to decline as the capitalist mode of production was consolidated in the countries that were the basis of its existence. The subordinating of merchant capital to industrial, natural to the new mode of production, was also manifested then in the realm of international trade. As a result the material conditions were laid for an ever increasing intensification of the action of the mechanism of the economic exploitation of all the countries of the world by a handful of 'advanced' countries. And the wider the gap in levels of labour productivity becomes between this handful of advanced countries and the economically backward ones, the greater is the exploitation suffered by the latter on the world market. This pattern was expressed with maximum clarity in the following formula of Marx:

The favoured country recovers more labour in exchange for less labour, although this difference, this excess is pocketed, as in any exchange between labour and capital, by a certain class.

In that way, in particular, the main antagonistic contradiction of the capitalist mode of production, i.e. that between its constantly increasing social character and the private capitalist forms of appropriating the results of social labour, thus found expression within the world market.

The putting of capitalism onto the rails of large-scale machine industry during the industrial revolution, which was completed in the second half of the nineteenth century, meant the beginning of a new, higher stage in the development of world economic relations. Large-scale industry in fact shaped the world market of the epoch of mature capitalism. And its dependence on external economic links and the international division of labour in turn also grew immensely. As Karl Marx said in his day: Large-scale industry, detached from the national soil, depends entirely on the world market, on international exchange, on an international division of labour.

In this situation the antagonistic contradictions of the world market stemming from the basic economic contradiction of capitalist society, have become an ever more serious obstacle to a further rise of social production. One of the most convincing bits of evidence of that were the world economic crises that were converted into a constitutionally inalienable feature of the whole economic system of capitalism. Years of comparatively rapid industrial boom began to alternate with regular consistency with periods of stagnation and direct destruction of already created productive forces. This cyclicity, determined by the objective laws of capitalist production, was intensified in turn by industry's mounting dependence on the foreign market. It was their comprehensive study both of the internal, home factors of the cyclic development of capitalist business and of the international ones that enabled the authors of classical Marxist political economy to draw the conclusion, confirmed by the whole subsequent course of social development, that economic crises would remain, while capitalism existed, one of the incurable ills determining its historically transient character.

For more than a century and a half now, since 1825, capitalism has periodically suffered from cyclic crises. All the attempts of capitalist economists to work up prescriptions to cure this illness without eliminating its basis, namely its exploiter relations of production, have unfailingly proved unsound and bankrupt. The capitalists as a class have demonstrated more and more clearly with the course of time their incapacity to control the productive forces of society rationally not only within the context of the separate national economies but also and especially on the scale of the world economy.

The world cyclic crisis of 1974–75, the greatest since the war, which developed under an unprecedented sharpening of several of modern capitalism's most important world problems, viz. the monetary, energy, raw material, and food crises, was convincing evidence of that indisputable fact in our day. It swept almost all the main centres of capitalism simultaneously and in turn interrupted their economic development. During the crisis the monopoly capitalists as usual exerted immense efforts to shift its most serious consequences onto the working masses of their own countries and the peoples of the agrarian and primary producer periphery.

As capital internationalises society's productive powers the connection between the reproductive processes of the individual countries (including those that are the epicentres of world crises under the impact of the economy's cyclic movement) is becoming closer and closer. Therefore, when we are determining the objective patterns of capitalism's development, the question of the interaction of the national cycles within the context of a single world cycle (study of which was begun by Karl Marx) is posed in all its acuteness. We shall dwell in detail on the features of the cyclic development of today's world capitalist economy later, here it is necessary to stress the following from the standpoint of the matter we are examining.

Despite the fact that the cyclic synchronousness of the movement of capitalist production and of the world market already lias a history of 150 years, a line of refusing to recognise the general patterns of this movement still clearly dominates capitalist economics. When, however, the course of the world cycle is studied, its development is mainly considered as a kind of arithmetical sum total of the results of cyclic fluctuations within the national economies taken separately. As a result it comes about that the specilic features of national development are, as it were, the sole decisive factor in the shaping of each world cycle. In that case the role of the system of international socio-economic connections and of the internationalisation of social production, which in essence convert these countries into links in a single economic chain, is patently underestimated.

Since the beginning of the 70s, it is true, because of the steady exacerbation of modern capitalism's international economic problems, an increasingly negative attitude toward sucli a 'narrowly nationalist' approach to analysis of the main world crisis processes (including its cyclic development) has been traceable in capitalist economic literature. Sometimes, while justly noting the close connection between these processes, and in particular between national overproduction crises, within the context of the capitalist economy, most capitalist economists as usual, however, prefer to identify their prime causes with the processes taking place at any definite moment in the sphere of the international market and political relations. They thus willy-nilly leave aside objective study of the system of international exploiter relations of production that underlies the deep internal contradictions of the world capitalist economy, which are inexorably undermining its main foundations.

The political dependence of most of the countries of the world on the main centres of capitalism has played an extremely important z'ole in the long history of the building up of this system. The structure of its unequal, exploiter relations was also moulded by that dependence in the imperialist stage, the crisis of which developed with particular force in the second half of this century as a consequence of most colonial countries' winning of national independence.

The Role of the Colonial System in the History of the World Capitalist Economy

The forming and rise of the capitalist mode of production was inseparable from the shaping of the capitalist world system of colonial oppression and unrestrained economic exploitation of some countries by others. The subsequent decline of the capitalist social system was intertwined in turn with the collapse of the colonial powers' political domination, break-up of the colonial system, and the rapid development of a neocdlonial policy by the former metropolitan countries as the crisis of the world capitalist economy deepened.

Relations of dominance and subordination on an international scale had long existed, of course, before the development of the capitalist social system, but the colonial policy of the successive exploiter formations was not identical in its essence and motives. It reflected the differences in principle of the socio-economic nature of the slave-owning, feudal, and capitalist systems. Under the slave-owning and feudal systems the conquest and colonisation of other countries did not in essence disrupt the economic basis of the life of their peoples, or the social structure of their society, and did not lead to any cardinal shifts in the structure of production in the conquered areas.

The most important feature of the colonial expansion of the leading European powers, already when capitalism was being born and winning against feudalism, was a policy of breaking down and liquidating the economic independence of the colonised countries, converting them into economic appendages of the metropolitan countries. The capitalist class reorganised the world in its own image and likeness. The European powers' colonial conquests, which were determined from the time of the great geographical discoveries by an ineradicable striving to capitalise profits, began to take on a new class content, and reflected the economic interests of the rising capitalist class.

However the ideologists of capitalism now try to twist and idealise the reasons for its colonial policy, and to separate its centuries long history from that of the forming and development of the capitalist mode of production, they are unable to disprove the conclusionunimpeachably demonstrated by Marxist political economy that the system of the enslavement and exploitation of some countries by others was the creation of capitalism itself. This system, which took shape during the struggle to build colonial empires, immensely promoted primitive accumulation of capital.

The idea behind the capitalists' colonial policy from the very start was that the material wealth expropriated and appropriated by them should he converted into capital only after it had been brought to the metropolis. For the pillaged countries it was nothing, in practice, but pure, uncompensated loss; not only was their economic organism bled white, but in many cases their productive forces were directly destroyed.

The system of international economic relations thus built up favoured the rise of the capitalist mode of production in only relatively few countries at first. As a rule the expansionist policy of the capitalist class of the metropolitan countries entailed economic regression in other areas of the world, disruption of the normal course of historical development, and destruction of the shoots of the new relations of production that had arisen there before the coming of the colonialists. That was one of the objectively inevitable patterns of the dialectic of the forming of capitalism's world economy as a system of exploitation of an immense mass of peoples backward in their social development by capitalist countries.

The several centuries long process of the formation of the world economy was only completed at the turn of the nineteenth and twentieth centuries when the whole world was shared out among a handful of capitalist countries.

The simultaneity of the completion of the forming of this economy and the territorial carving up of the world (like the simultaneity of the birth of these processes) was quite natural and regular. But the coincidence in time is also due to a certain vagueness and confusion in the definition of the relation of such very important concepts as the 'colonial system' and the 'world economy'.

In a number of works by Soviet and other Marxist economists, especially in the 30s and 4Us, the world capitalist economy was not treated as a special category of political economy and object of special investigation. It was primarily studied as a consequence of the colonial policy of the imperialist powers.

This approach was also reflected iu Soviet textbooks of the time, in which the term 'world capitalist economy' was often employed mainly in connection with formulation of the territorial division of the world as one of the main economic attributes of imperialism (its main features usually being treated, moreover, in chapters, devoted to description of the colonial system).

The collapse of the colonial system in the postwar decades graphically demonstrated the illegitimacy of such an approach. The world capitalist economy continued to exist as well with the break-up of the colonial empires, and to function as a mechanism of the economic exploitation of some countries by others, and as a system uniting the countries of the non-socialist world in a single economic organism in which the international relations of the capitalist mode of production prevailed.

In the light of the new facts it naturally became necessary to get a thorough understanding, methodologically, of the interdependence and relationship of these categories. It was not fortuitous, therefore, that by the end of the 50s special sections and themes had already begun to appear in certain studies of the international economic problems of imperialism that treated the significance of the colonial system as one of the most important, but far from determinant, elements of the postwar capitalist economy.

But because of the subsequent completion of the break-up of the colonial system evaluations of that kind also required amendment. The collapse of the last great colonial empire, the Portuguese, at the beginning of the 70s, signified that 33 the world capitalist economy had irrevocably lost one of its former colonial foundations. Its periphery now consists almost wholly of sovereign national states and no longer of rightless colonies and semi-colonies.

As a result, the territorial (political) division of the world has now finally lost its former signiiicance. Its other characteristics, which ultimately governed the development of a broad system of neocolonial dependence of the emancipated countries on the former metropolises, has taken over the dominant role in imperialism's expansionist policy.

These characteristics include the struggle for an economic partition of the world and for the export of capital, which were directly linked with the inter-imperialist struggle for a territorial division of the world throughout the world capitalist economy's preceding colonial history.

Many of the forms and methods of neocolonialism stemming from the export of capital and the struggle of monopolistic alliances for an economic partition of the world by no means came about suddenly, and have not simply been the consequence of the hopeless crisis of monopoly capital's traditional colonial policy after World War II.

Their sources lie deep in the period when the strategy of territorial conquest and colonial expansion held undivided sway in world economics and politics. As this expansion developed the role of capital exports and of the imperialist struggle for an economic division of the world gradually rose.

The significance of the system of political enslavement itself, moreover, as the cementing force uniting extremely heterogeneous countries and territories (as regards level of development), during the moulding of imperialism's world economic relations into a single economic whole, did not remain static.

The colonial system played a leading role in this respect, above all iu the early stages of the formation and development of these relations.

But as more and more active forms of dependence of colonies and semi-colonies on the metropolitan countries, and on the whole of backward agrarian countries on industrially developed ones, built up thanks to it, those forms and methods of economic enslavement came to the fore that later also underlay imperialism' s neocoloriialist strategy.

In the concluding stage of the formation of the capitalist economy, when large-scale machine industry developed 34
rapidly within it, the colonial system began to perform a new function compared with the preceding stage, and fostered the drawing of the colonies and semi-colonies into the international division of labour built up by monopoly capital on the basis of this industry.

The unprecedented extension in the scale of monopoly capital's external economic expansion led to a final consolidation of both politically rightless countries and sovereign ones in the role of primary commodity appendages of its industrial centres. It was then that the conditions for the economic exploitation of agrarian countries and raw material suppliers, without their direct political enslavement, began to build up particularly rapidly. Under imperialism, Lenin stressed, economic 'annexation' is *fully* 'achievable' without political annexation and is widely practised.

At the beginning of the twentieth century there were no more 'free' countries, i.e. lands not exploited by the handful of imperialist powers. In those conditions the inter–imperialist struggle for dependent and semi-colonial countries became extremely tense, and the greatest progress was made in it by the United States of America, the economically strongest capitalist power.

The infinitely growing significance of the export of capital from the metropolitan countries to their primary commodity peripheries predetermined the development also of another feature most essential for a description of the world economy in its monopoly stage, and that was the tendency to an acceleration of the capitalist development in economically 35 backward countries. After World War II, as the hulk of the colonies and semi-colonies achieved national sovereignty, this tendency (as will he shown in the following chapters) acquired new, original forms.

The postwar break-up of the colonial empires had a substantial impact on the development of several of the decisive contradictions of the imperialist economy, lint one will by no means get a full idea of the specific nature of those contradictions if one does not take the colonial past into account. Analysis of the long-term trends in the reciprocally conditioned development of the capitalist mode of production, the world market, and the colonial system enables one better to understand and comprehensively evaluate the substance of the structural shifts that have taken place in the world economy, given the steady deepening of the general crisis of capitalism.

4

History of the World Capitalist Economy

INTRODUCTION

The history of the world economic system oi capitalism is inseparable from that of its conversion into a dominant economic formation whose structure is considered by Marxist-Leninist economics as a living, constantly developing organism and not as something mechanically interlocked and therefore admitting of all kinds of arbitrary combinations of the individual social elements. Study of its current trends of development and of its antagonistic contradictions necessarily calls for a determination (albeit very cursory) of the main stages of the passage of international capitalist relations of production from one qualitative state to another.

The objective preconditions of the deepening crisis and decline of the previously all-embracing system of the world capitalist economy were built into the 'genetic code' of capitalism itself. The detailed theoretical analysis of them made in their day by the founders of Marxist-Leninist political economy helps bring out both the general patterns and the specilic features of its long-term line of development in recent decades.

The Pre-monopoly Stage of Capitalism

Every new mode of production replacing a preceding one regularly creates broader opportunities for further growtli of the productive forces. The capitalist system made a great stride forward compared with feudalism opening up hitherto unprecedented perspective for the development and internationalisation of social production. The decisive factor marking the start of the capitalist epoch was the decomposition of feudal relations of production and the rise in several 37 European countries from the middle of the sixteenth century of the foundations of capitalist production in the form of various types of manufacture. As a result an economic community of separate countries took shape, and a system of division of labour developed on a national and an international scale, home markets were formed, and exchange grew between them.

The gradual consolidation of the economic positions of capitalism (international ones included) prepared the material, and consequently the socio-political, conditions for bourgeois revolutions in a number of countries in Western Europe. These revolutions, by overthrowing the political superstructure, in turn accelerated the replacement of feudal relations by capitalist ones; at the same time they cleared the road for steady external economic expansion of the European capitalist class.

The thirst to enrich themselves at the expense of other nations constitutionally inherent in all exploiter formations became an urgent economic need in the course of rapidly developing commodity production. It determined the capitalists' need to open up trade routes from Europe to Asia, Africa, and America. The great geographical discoveries were the natural consequence of the unprecedented growth of commodity and money relations in European countries evoked by the development of capitalist production and the corresponding expansion of the national and world markets.

The merchant capitalists had a paramount role in all those processes. The East India Companies—the Dutch (1602–1798), English (1600–1858) and French (1664–1770, 1785–1793)—wrote bloody pages in the history of capitalism. These great joint-stock trading organisations, with a capital very considerable for their day, promoted the drawing of colonised countries into international trade in every possible way, and converted them into economic appendages of the metropolitan countries.

The formation of capitalism's world socio-economic system thus took two main directions from the very beginning: (1) along the line of development 'in depth', i.e. through the growth of capitalist production and national markets in a comparatively small number of economically advanced countries; (2) along the line of development 'in breadth', above all through expansion of a handful of capitalist countries' colonial domination to newer and newer countries 38 and territories, and the forging of stronger interconnections in the world market.

Capitalism, being a natural stage of historical progress, and raising labour productivity to a higher level, thus inevitably broke up relations of production based on personal dependence. But the capitalist society growing from the womb of feudalism did not eliminate class contradictions; on the contrary, it fostered their further exacerbation both within separate countries and on the international arena. Capitalist production only replaced old classes, old conditions of oppression, and old forms of struggle with new ones. As capitalist manufacture grew an urgent need matured to pass to a larger scale, factory production, which ultimately led to the industrial revolution. Developing in the second half of the eighteenth century, first in Great Britain and later embracing a number of other countries, this revolution was completed about the middle of the nineteenth century. The genesis of the factory system of production employing machine technology, and the vast extension on that basis of the international sphere of the capitalists' class omnipotence through colonial expansion, also meant confirmation of capitalism as the prevailing mode of production.

The changes taking place in the productive forces of that time under the action of capitalism were really impressive. In England between 1800 and 1850 industrial consumption of cotton rose more than 14-fold, the mining of coal over fivefold, the smelting of iron by around 12-fold. In France, in the second quarter of the nineteenth century alone, output of coal almost trebled, and the production of iron doubled. The production of industrial goods increased at unprecedented rates in Germany, the USA, and certain other countries. At the same time capitalism's international 39 commercial and industrial relations developed rapidly. Because of the application of steam engines in transport there were qualitative shifts in the international lines of communication that linked the overwhelming majority of countries together at that time. World trade rose by nearly 180 per cent in the first half of the nineteenth century.

These changes, however, so impressive for their contemporaries, now seem quite modest. In fact they only prepared the ground for a subsequent rapid increase of capitalist production.

Yet it was the transition from manufacture to largescale mechanised industry, accompanied with the development of fundamentally new industries, and the drawing of additional manpower and material resources into production, that ultimately led to formation of the material and technical basis of pre-monopoly capitalism. The essence of the process was succinctly characterised by Lenin:

> The progressive historical role of capitalism may be summed up in two brief propositions: increase in the productive forces of social labour, and the socialisation of that labour.

Capital could not perform that historical function without internationalising the productive forces and without a struggle to mould a corresponding world economic system. The building of that system, which caused a gradual economic convergence of nations, and a strengthening of their economic relations and of the international division of labour, was an inevitable manifestation of the natural.process of capital's socialisation of labour.

That process, however, did not simply lead to consolidation of the bourgeoisie's class position; its most important consequence was that it objectively prepared the national and international socio-economic conditions for the inevitable transition from capitalism to the next, more advanced social formation which by completely abolishing exploiter relations would open up unlimited scope for a further increase of humanity's productive powers.

Together with growth of capitalism in separate countries, and extensioji of its international influence on the world arena, the working class came forward as the main driving force of further historical progress. As Marx wrote:

> The bourgeois period of history has to create the material basis of tho new world—on the one hand universal intercourse founded upon the mutual dependency of mankind, and the means of that intercourse; on the other hand the development of the productive powers of man and tho transformation of

material production into a scientific domination of natural agencies. Bourgeois industry and commerce create these material conditions of a new world in the same way as geological revolutions have created the surface of the earth.

The growth rates of capitalism's productive forces on the basis of large-scale production increased markedly in its main centres in the second half of the nineteenth century. In 1880 the volume of industrial production had almost trebled for the whole world compared with 1850. Important shifts had occurred, or were in the bud, in the very structure of capitalist production as a result of faster growth of the main branches of heavy industry. Iron and steel, mechanical engineering, and the chemical industry began to be converted into the leading branches of the economy. The system of international division of labour continued to expand rapidly. And the development of transport, in which fundamentally new technique was intensively introduced, played a very essential role in cementing capitalism into a single economic whole.

Pre-monopoly capitalism reached its highest stage by the last quarter of the nineteenth century. Then a stormy process of the rise and growth of monopoly enterprises of a jointstock type—cartels, trusts, and syndicates—began in its main centres. Capitalism began to grow over, as a world social system, into its highest and last stage of historical development.

The Monopoly Stage of Capitalism

The building of an all-embracing imperialist system was completed at the turn of the century, when the undivided sway of monopoly capital was established in the main capitalist countries, and the territorial division of the world between them was fully completed. The main patterns of the functioning of monopoly capitalism began then to determine the decisive lines of development of the world economy and world politics. The concluding stage of the formation of a world economy, which took around two or three decades, belongs roughly to the same period.

The transition irom pro-monopoly capitalism to imperialism was I he result of deep-seated shifts, above all in the productive forces and technical basis of society, under the action of the ever accelerating concentration and centralisation of production and capital in its main centres. The development of monopolies inevitably retarded technical progress, but at tho same time the increasing concentration of their economic and financial power opened up hitherto unprecedented opportunities for broader application of those advances of science and engineering in the world economy that helped them increase their profits.

This period was marked by the introduction, especially in heavy industry, of major technical advances that brought into being economically more efficient forms and methods of production. The competitiveness of the monopoly mergers was enhanced and the objective conditions created for a further acceleration of the concentration and integration of production. The technical advances that had a marked effect in changing the structure and accelerating growth of the world capitalist economy included, for example, the invention of new ways of making steel (the Bessemer process in 1855, open hearth process in 1864, and the Thomas, or basic process in 1878). Their introduction by the monopolies of the main capitalist countries made it possible to convert the iron and steel industry into a leading branch of the world economy. Whereas only around 500,000 tons of steel was produced in 1870 in the whole world, production was million tons in 1890, and already 76 million tons in 1913. World production of pig iron rose to 24 million tons in 1890 and to nearly 80 million tons in 1913.

The development of the engineering and chemical industries also had a considerable impact on the advance of many other branches of the economy. The application of technical advances in production that led to the perfecting of prime movers played a revolutionary role in tin's respect. Among them were the invention of the dynamo (1867), the fonr-stroko internal combustion engine (1877), the steam turbine (rnid-80s), and the Diesel engine (end of the 90s). The new types of engine promoted the development of new forms of transport: the motor car (1885), the Diesel locomotive (1891), and the aeroplane (1903). The application of improved means of transport and communication, which had in fact united all areas and countries of the globe at the beginning of the twentieth century, immeasurably accelerated the forming of the world system of division of labour of the imperialist epoch. In 1913 the length of the world's railways was more than a million kilometres, of which more than 250,000 km had been built in colonial and dependent countries in Asia, Africa, and Latin America. There was also a steep increase in marine freights and ocean transport. The energy base of world capitalism, then based on coal, also grew rapidly. The mining of coal exceeded 1,000 million tons at the beginning of this century. The technical basis of the power industry also broadened considerably. For many decades after Watt's invention of the steam engine capitalism's economic growth was linked with the application of steam, but at the turn of the century the use of electricity was becoming more and more important, which also fostered the introduction of a number of new advances in the field of electrical engineering by the biggest companies: viz., the electric smelting of steel (1877), the electrolytic technique of obtaining non-ferrous metals (end of the 1880s), longdistance transmission of electricity (1891), etc.

The world level of production in manufacturing industry rose by 120 per cent between 1880 and 1900, and by 270 per cent by 1913. World trade in manufactured goods rose by 210 per cent (in constant prices) in the same period, and in primary commodities by 220 cent. By the beginning of this century the lion's share of world industrial output and trade was already concentrated in the hands, or under the full control, of the finance capital of ten imperialist powers, which led in a comparatively short time to establishment of the unlimited domination of imperialist monopolies in international economic and political relations. All that in turn brought about essential modifications in the pattern of development of the world economy compared with the period of free competition. Since all countries and nations without exception had been drawn into the orbit of imperialist domination, elucidation of this pattern called urgently for theoretical comprehension of the actual changes taking place both in the economies of the main capitalist powers and on a world scale.

It was necessary to get an overall picture of the whole capitalist economy and of the most important trends in its development. Lenin took it on himself to tackle this problem. The concrete, historical analysis of the main tendencies of economic development at the beginning of the century that he made in his book *Imperialism, the Highest Stage of Capitalism* enabled him to define imperialism as a special stage of capitalism within which monopolies and finance capital had established world domination, the export of capital had acquired salient significance, economic division of the world between alliances of capitalists had begun, and the territorial partition of the world among the capitalist powers had been completed.

This definition of the imperialist stage included its main economic attributes, which at the same time also characterised other very important aspects of the affairs of capitalist society at that time in the realm of politics, social relations, class struggle, ideology, and culture. It was a matter, consequently, of a single chain of socio-economic and political changes that brought to light the very essence of the monopolistic stage of capitalism as a world system rather than of isolated or one-sided attributes In defining this essence, Lenin noted that on the one hand, finance capital is tho bank capital of a low very big monopolist banks, merged with tho capital of tho monopolist associations of industrialists; and, on the other hand, the division of tho world in tho transition from a colonial policy which has extended without hindrance to territories unseized by any capitalist power, to a colonial policy of monopolist possession of tho territory of the world, which has been completely divided up.

The various fornib of monopoly in one country or another, and in the various branches of the capitalist economy had already arisen on the national and world markets, as we know, under free competition, but they only acquired a dominant position on a world scale under imperialism. Marx and Engels, in wringing out the deep-sealed tendencies of development and the historical inevitability of the destruction of the capitalist formation, had been able in their day to trace the genesis of capitalist society from its embryonic state to the period when the joint-stock form of monopoly capital began to arise within it.

They could not then foresee, of course, that capitalism would pass to a higher stage of development, but they perceived, with the shrewdness of genius, in the joint-stock form of the 'self-movement' of capital a tendency toward monopolisation that sapped the very principle of free competition. At the same time, their studies showed that this tendency, reflecting natural growth of the social character of capitalist production, must promote the creation of new objective conditions for mankind's revolutionary passage to a more progressive social system.

Lenin's theory of imperialism brought together a Marxist analysis of the patterns of capitalism and a study of the qualitatively new processes and phenomena of world development at the beginning of the twentieth century. Elucidation of the economic essence of the new stage of capitalism was the core of this theory. Just as free enterprise was the main trend in the affairs of the world capitalist economy in its early stages of development, so establishment of the domination of finance capital on a world scale was the most important feature of its imperialist stage.

Because the economic basis of capitalism consists in capitalist private ownership of the means of production, a uniting of two contradictory principles—monopoly and competition—is typical of the world economy of imperialism. At the same time it retains a broad spectrum of both pre-monopoly and even pra-capitalist forms of commodity production. Without them it would be inconceivable, since imperialism, as Lenin often stressed, is a kind of superstructure on the old capitalism that is the admitted consequence of the development, expansion, and continuation of the deepest, most fundamental tendencies of capitalism and commodity production in general. That formulation of the problems acquires special significance when we analyse the 45 structural changes in the world capitalist economy with the collapse of its colonial basis.

The national features of the monopolisation, and consequently, too, of the internalionalisation, of large-scale proel uction and capital in no way rule out the operation in each country of the general patterns that in the aggregate determine the unity of the various features of imperialism's economy and politics. Not all the main characteristics of the highest and last stage of capitalism, moreover, are manifested in each country to the same exten! and in all areas. The possession of colonies, for example, or the export of capital, played a minor role, or did not occur, in the specific conditions of its formation in certain imperialist countries, but the existence of the decisive characteristics—the domination of monopolies and finance capital—was obligatory. The economic attributes of imperialism, taken together, are manifested in their unity in the world capitalist economy.

At the turn of the century this economy was wholly transformed into an antagonistic aggregate of two main groups of countries, imperialist countries on the one hand and the colonies and dependent countries exploited by them on the other hand. In that connection we must remember that some of these characteristics had already begun to be observable in Great Britain in the middle of the nineteenth century. By the end of the century, however, Britain's advantages had largely been forfeited. One country's domination had given way to the monopoly position of several main, competing imperialist powers.

Although the policy of colonial enslavement of other countries in the capitalist system historically became very common, imperialism had nevertheless begun, right at the start, as we showed above, to create qualitatively new conditions for the expansion of finance capital in economically backward countries without their direct political subjugation. The patterns of the division of labour between developed capitalist, countries and economically backward agrarian ones 46 operating ai that time favoured further consolidation of these conditions.

The omnipotence of finance capital in world economics and politics had grown to such an extent then as to enable the leading capitalist countries to convert even individual colonial powers that were lagging in economic development into dependent countries. To some extent this dominance enabled many features of the former structure of the uneven division of labour in the world capitalist economy to be retained subsequently, after the disintegration of the colonial system.

One of the most essential features of world economic relations in the monopoly stage of capitalism is the export of capital. By greatly intensifying the foreign economic expansion of the main capitalist countries, it enabled the monopolies of the imperialist powers to gain a predominant influence in other countries, often with the letter's retention of actual national independence. It also led to the development of the feature noted above, namely the tendency to an accelerated but dependent development of capitalism in colonies and other economically backward countries. The colonial policy of the monopoly capitalists continued, at the same time, to pursue the goal of maintaining survivals of pre-capitalist social structures in agrarian countries and raw-material suppliers (especially as regards their agrarian relations) as a social prop for their dominance and source of extra superprofit. The intertwining of pre-capitalist forms of production with monopoly forms and methods of exploitation intensified the contradictions of the world economic system, and led to a further increase in tlie unevenness of the development, of its various parts, and of separate countries and regions.

At the same time monopoly capital added to the old methods of capitalists' expansionist policy a struggle for redivision of the already divided world, to take up its raw material resources and profitable spheres for the employment of capital, for new potential markets for their industrial output, and for spheres of monopoly control over economically exploitable areas in general. The inter-imperialist struggle fostered a steep increase in the spasmodic character of the economic development of various countries, which became a most important feature of the functioning of the world capitalist economy over the whole of its subsequent history.

With the transition to the monopoly stage there were also considerable changes in the political and ideological superstructures of capitalist society. The determinant moment of all these changes was a strengthening of reaction all along the line. A period of unrestrained intensification of exploitation and national oppression of all the peoples of the world by the monopoly capitalist powers set in, a period of imperialist struggle for spheres of complete bossing of the show in the world. In these conditions the capitalists as a class finally lost their once progressive role.

In creatively developing the Marxist theory of social development on the basis of a thorough study of this stage of capitalism, Lenin came to the very important conclusion that a fundamentally new historical situation had arisen promoting a further upsurge of the world revolutionary process:

Capitalism has become reactionary; it has developed the forces of production to such a degree that mankind is faced with the alternative of adopting socialism or of experiencing years and oven decades of armed struggle between the 'Great' Powers for the artificial preservation of capitalism by means of colonies, monopolies, privileges and national oppression of every kind.

The unprecedented sharpening of the contradictions of capitalist development on a global scale created the objective conditions for snapping the single chain of imperialism at its weakest links, initially in a few countries, or one, taken by itself. Imperialism's historical place in relation to capitalism in general was thus defined as a turning point in society's affairs, as the eve of a new historical epoch of revolutionary suppression of the capitalist social formation by a socialist one.

The World Capitalist Economy Today

The First World War (1914–1918), and the victorious socialist revolution that developed at the end of it in Russia, opened a new chapter in world history. Imperialism's previously all-embracing political and economic system began to break up. Over an area of one-sixth of the world's land surface, with a population then of 120 million, there arose and began to develop a fundamentally new social system. The productive forces of socialism began to be made an integral component of mankind's productive forces.

As a result of the formation of the first socialist state, capitalism could no longer dominate the world economy without limit, as before. A historically inevitable period of coexistence, struggle, and interaction began in the world economy and in world politics between two socio-economic systems based on fundamentally opposite modes of production.

It is well known that imperialism exerted immense efforts in order to block this course of events. After the repeated failure of attempts to strangle the socialist system at birth by force of arms, the monopoly circles of the West took the line of economically and politically isolating the young Soviet Republic, whose economy had suffered enormously from civil war and foreign intervention.

The monopoly capitalists thus endeavoured to reverse the processes in the world economy that expressed the vital needs of a further advance of society's productive forces, but these plans to counteract the objective economic patterns of social development disclosed by Marxist-Leninist political economy also suffered one failure after another. As Lenin remarked:

There is a force more powerful than the wishes, the will and the decisions of any of the governments or classes that are hostile to us. That force is world general economic relations, which compel them to make contact with us.

The forces of world reaction subsequently strove incessantly to put obstacles in the way of the building of a system of mutually advantageous international economic co–operation of countries with different socio-economic systems. The story of the moulding of this system has been one of consistent, persistent struggle by socialism and all mankind's progressive forces to preserve a world division of labour and develop it on a new basis, and to establish truly equal commercial and economic, scientific and technical, and cultural relations between all countries. Success would have been impossible in that struggle if it had not corresponded to the real interests of nations and of the non-socialist countries, and also if it had not met the requirements of the objectively operating laws of social development, and if it had given a one-sided advantage to the socialist countries. Thus, even between the wars, the basis was laid for intricate, contradictory processes determining the most vital directions of the subsequent crisis development of capitalism's world economic system.

After World War I there was a comparatively rapid restoration and further expansion of production, and of the world economic links broken by the war, within contracted boundaries. But this growth took place on a background of further deepening of imperialism's basic contradictions, a sapping of its colonial foundations, and an intensification of the unevenness in the development of separate regions and countries.

At the same time the socialist mode of production was more and more clearly becoming a constituent part of the global economy of mankind, and was graphically demonstrating its vitality. The international standing of the Soviet Union was strengthened, and its external economic 51 relations extended, through intergovernmental agreements with several capitalist conn tries, certain former colonies and semi-colonies. The progress made in building socialism in the USSR, and the consolidation of its position on the international arena, where the foundations of a new world social system were being laid, became vital factors in the subsequent aggravation of the general crisis of capitalism.

Growth of capitalism's irreconcilable contradictions continued to play the determinant role in this process. Particularly convincing evidence of that was the world cyclical overproduction crisis of 1929–33, the greatest in scope, duration, and disruptive consequences in capitalism's history, which simultaneously swept the industrially developed countries and their agrarian, raw material periphery. The crisis, which threw the economy of the capitalist world several years back, had a very negative effect as well on the whole system of imperialism's global economic connections. By the end of the 30s the volume of international trade, and also of capitalist industrial production was below the level of 1913. Right up to the beginning of World War the economies of most countries of the capitalist world had still not been able to recover fully from that crisis.

At the same time the colonial foundations of the capitalist economy had- been markedly weakened. The successful anti-imperialist struggle of the peoples of the USSR and their achievements on the road of radical socio-economic reforms, stimulated an upsurge of the national liberation movement in many colonies and dependent countries, some of which had managed after World War 1 to win or retain their national independence. A crisis of the all-embracing colonial system began. The colonial powers, it is true, were then still able to avert the break-up of their empires. Without ceasing their struggle to redivide spheres of domination, they endeavoured as before to deprive the countries economically dependent on them of national sovereignty; and to that end imperialism repeatedly unleashed predatory wars (e.g. Italy's aggression against Ethiopia, and Japan's against China).

Characteristically, however, not a single imperialist power could succeed, even in the initial stage of the general crisis of capitalism, in turning any economically backward country into a new colony. While retaining a decisive position in the world economy, linance capital was forced to take into account in its expansionist policy not only the 'strength' and 'capital' of its imperialist rivals but also of the growing strength of resistance of the peoples of the countries exploited by them, who received every possible support from the iirst socialist country.

World War II, which broke out as a result of the sharp aggravation of all capitalism's contradictions at the end of the 30s, and of the striving of its most aggressive circles to strike a crushing blow against socialism and the national liberation movement, in the final count led to a further deepening of the crisis of the imperialist system, the boundaries of which contracted even further. The socialist mode of production became predominant in a number of countries in Europe and Asia, and a new phase opened in the economic competition of socialism and capitalism.

The antagonism and interaction of the two opposing social systems became a determining feature of today's world economy in the postwar years; in that connection it has become necessary, when studying the structural shifts and fundamental features of the postwar development of capitalism, to take into account at the same time the global tendencies of the growth of production, the more so that the long-term consequences of the scientific and technical revolution that has embraced the leading spheres of the productive activity of human society as a whole are beginning to have a mounting effect on both socialist and capitalist countries, and also on many developing ones.

Analysis of the facts characterising these tendencies indicates that in the period under review mankind's productive forces have attained a height hitherto unprecedented. The international exchange of scientific and technical knowledge and production know-how has reached a fundamentally new level. The infrastructure of world economic relations has also developed rapidly under the impact of the scientific and technical revolution on the basis of a national and international division of labour. On the whole, the volume of the aggregate product of all the countries of the world was more than four times as big at the end of the 70s as at the beginning of the 50s. There was also a marked increase in the main productive forces, viz., manpower (the 53 number of the world's inhabitants rose by almost 75 per cen! in the same period).

The list of that kind of quantitative indicators could, of course, be considerably extended, but the figures cited give a graphic idea of the sizable shifts that have occurred in humanity's productive forces in the contemporary historical situation of competition of the two main modes of production and the corresponding two world socio-economic systems, namely the socialist and the capitalist.

As Karl Marx remarked in a letter to P. V. Annenkov (28 December 1846), tho productive forces are therefore the result of practically applied human energy; but this energy is itself conditioned by the circumstances in which men find themselves.

Which of the opposing world social systems has the advantage in creating conditions for growth of the productive forces can be judged from the following facts. On the whole, in the three decades of the 50s through to 80s, the rates of growth of national income and industrial output in the socialist countries belonging to the Council for Mutual Economic Assistance (CMEA) were three time as high as *in* all the capitalistically developed countries taken together. The countries belonging to CMEA are the most dynamically developing economic community in the world. At the end of the 70s these countries, which have 10.2 per cent of tho world population, produced around one-third of the world's industrial output. In 1968–79 alone the volume of the gross social product in the European members of CMEA increased more than fourfold, while that of all the developed capitalist countries increased by only 140 per cent. A natural consequence of the development of this tendency has been a considerable raising of the role of socialism in the world economy. At the beginning of 1980 the population of 54 countries not yet broken free from direct, colonial dependence was not more than 0.1 per cent of all mankind.

The turn of the decade to the 80s was marked by further growth of the economic might of the countries of world socialism. As the Central Committee's report to the 26th Congress of the Communist Party of the Soviet Union noted:

The past few years have not, been among the most favourable for the national economies of some socialist states. Still, in the past ten years the economic growth rates of the

CMEA countries have boen twice those of the developed capitalist countries. Tho CMEA members continued to bo the most dynamically developing group of countries in the world.

At the same time the break-up of the colonial system was not limited to emancipation of the colonies, but also included semi-colonial and dependent countries. Throughout the long period ofthe imperialist powers' unlimited sway in the world economy the dominant tendency was to convert the latter into rightless colonies; in the postwar years another trend began to predominate, that of converting them into politically independent national states. Imperialism could not suspend the development of this tendency as well. In recent decades the overwhelming majority of semi-colonies and dependent countries, having thrown off foreign and proimperialist regimes, have also in fact broken free of imperialism's system politically, and are no longer its reserve, as they used to be. With the all-round support of the socialist countries and international working-class movement, they are intensifying the anti-imperialist struggle for economic independence. As a result there has been a marked quickening of the growth rates of their economies. The total production of goods and services in the developing countries of Asia, Africa, and Latin America rose more than 320 per cent in the 50s to 80s, industrial output rising by 620 per cent and farm production by nearly 130 per cent.

Their position in the world capitalist economy, however, remains extremely difficult and unequal. In spite of the fact that they have three-quarters of the population of the non-socialist world, they produced less than one-fifth of 55 that world's gross output at the beginning of the 80s. The immense gap in levels of development between them and the group of imperialist powers historically built up under capitalism has continued to widen. In the first postwar years, for example, this gap, in terms of per capita GDP, was characterised by a ratio of 1: 10, and at the beginning of the 80s by a ratio of more than 1: 13. In the same period there was a marked reduction of the weight of the former colonial world in the turnover of the world capitalist market.

The capitalist economic system is more and more clearly demonstrating its incapacity to deal with the pressing socioeconomic problems of emancipated countries. The logic of history is convincing their peoples of the lack of perspective of development along the capitalist road.

It is not fortuitous that many of the new national states, including some that are still largely enmeshed in imperialism's world economic relations, are rejecting the capitalist road of development in their constitutions, under pressure of the broad public, and are proclaiming a line of building a socialist society in the long term.

Leonid Brezhnev, characterising the significance of the countries liberated from colonial subjection that have taken the road of revolutionary, democratic transformations, and of countries with a socialist orientation in today's world, said that at the 26th Congress of the CPSU:

> Development along the progressive road is not, of course, the same from country to country, and proceeds in difficult conditions. But tho main lines are *similar*. These include gradual elimination of the positions of imperialist monopoly, of the local 56 big bourgeoisie and the feudal elements, and restriction of foreign capital. They include the securing by the people's state of commanding heights in the economy and transition to planned development of the productive forces, and encouragement of the cooperative movement in the countryside. They include enhancing the role of the working masses in social life, and gradually reinforcing the state apparatus with national personnel faithful to the people. They include anti-imperialist foreign policy. Revolutionary parties expressing the interests of the broad mass of the working people are growing stronger there.

Monopoly capitalism is really losing its old opportunity to decide the fate of nations even within the shrunken framework of the world capitalist economy. The main point in the people's national liberation struggle at the present stage of the general crisis of capitalism"is that it has been converted in practice into a fight against exploiter relations in general, both feudal and capitalist, in many countries. As Leonid Brezhnev has put it:

> It is of exceptional importance that many of the countries that have achieved liberation have rejected the capitalist road of development and adopted a socialist orientation, setting

themselves the goal of building a society free from exploitation. And that, without doubt, is a very strong blow to capitalism's positions as a whole as a world social system.

A substantial socio-economic consequence of the collapse of the colonial system is the relative lowering of the role of expatriate capital in the economies of emancipated countries. Before independence almost all their industrial production was in the hands of foreign monopolies and was part of the economy of the imperialist system. Today broad measures have been taken in'many of them both to nationalise expatriate enterprises and establish public control over their activity, and to found their own national industry.

There are few summarised, reliable facts in the international statistics on what proportion of industry remains under the direct control of expatriate monopolies, but varrious indirect indicators suggest that a considerable^part of the industrial potential of the developing countries, yielding more than 4 per cent "of gross world industrial output, has already passed *out* of such control.

Because of the effect of these tendencies, a further weakening of the imperialist powers' positions in the world economy has become typical of the period of the general crisis of capitalism, which finds expression above all in the obvious undermining of their former hegemony in world industrial production. At the end of the 30s the monopolies of the capitalist countries (including colonies and dependent countries) in practice controlled around nine-tenths of total world industrial production; at the beginning of the 80s only a little more than half of it was under their control.

The facts adduced here primarily bring out one of the decisive trends in the crisis of the imperialist economy, when the basic features and main peculiarities of contemporary world development are beginning to be determined by the working class and its offspring, the world socialist system, and no longer by the monopoly capitalists. The steady strengthening of the economic might of the socialist countries, the advance of the national economies of the new sovereign states, and the rapid development of mutually advantageous international economic relations between them are leading in turn to a marked limitation of the imperialist monopolies' diktat in world economic relations, including those between developing and developed capitalist countries on the world market.

Even the most persistent and subtle defenders and propagandists of the capitalist system are unable to deny the tendencies noted above, which are finding wide reflection in international statistics, and which express the main trends in the deepening of the contemporary crisis of the world capitalist economy. An ambivalent approach to these processes is therefore typical of capitalist politicians, economists, and sociologists, who are engaged in apologetics for imperialism. On the one hand, they prefer in every way to evade comparison of the long-term results of the development of the two world economic systems in their interconnections, and strive to avoid a comprehensive socio-economic analysis of these results. While the work of some leading capitalist scholars 58 undertakes to examine the dynamics of l4io economic growth of the opposing social systems today, it usually mentions the advances of socialism in passing as something irrelevant for an evaluation of the present-day level and the ensuing outlook for the global development of mankind's productive forces. On the other hand, immense attention is paid in Western capitalist literature to another aspect of this problem, namely, the factors that continue to promote growth of production within the capitalist economy. These factors are usually considered out of context of the processes showing overtaking growth rates in the economy of the world socialist system.

The ideologists of the capitalist system, glossing over its class and other antagonisms in every way, are relying on the development trends of social development in its industrial centres under the impact of the developing scientific and technical revolution. What lack of prospects can there he talk of, they ask, and even more what hopeless crisis of this social system, if capitalism's productive forces still continue to grow, albeit unstably and with sharp fluctuations The increase in the production capacity of the postwar world capitalist economy is thus advanced as the 'weightiest' argument to substantiate the 'historical stability' of capitalism and consequently, as well, the flimsiness of the Marxist-Leninist analysis of the irreconcilable contradictions of its development.

In fact the Marxist-Leninist theory disclosed another causal connection between the growth and internationalisation of social production under capitalism and its historical destinies, than that which its capitalist critics try to ascribe to it. The objective tendencies in the development of the capitalist mode of production are not leading to a stopping of growth of the productive forces, or to their stagnation, but on the contrary are leading to their accelerated advance compared with preceding formations. In the *Manifesto of the Communist Party* Marx and Engels wrote:

The bourgeoisie cannot exist without constantly revolutionising the instruments of production, and thereby the relations of production, and with them the whole relations of society—Constant revolutionising of production, uninterrupted disturbance of all social conditions, everlasting uncertainty and agitation distinguish tho bourgeois epoch i'rom all earlier ones.

But does all that mean that, as the productive forces of capitalist society develop, its economic basis is perpetuated and becomes ever firmer and more unshakeable? The founders of Marxism-Leninism, in bringing out the inner contradictory dialectic of that development, demonstrated that it in no way contributes in the end to the consolidation of capitalism. It relentlessly undermines the foundations of this system, and forms the conditions within it necessary, for victory of a more advanced social system. By accelerating growth of the productive forces in its own class interests and internationalising them in every way, the capitalist class itself prepares the material and social conditions for destruction of the social formation created by it.

The transition to monopoly capitalism, while heightening the unevenness and spasmodic character of its development, was marked by a further growth of capitalist production. In describing imperialism as the highest and last stage of capitalism, in which tendencies toward decay and growth of society's productive forces predominated simultaneously, Lenin stressed that:

It would bo a mistake to believe that this tendency to decay precludes the rapid growth of capitalism. It does not. In the epoch of imperialism, certain branches of industry, certain strata of tho bourgeoisie and certain countries betray, to a greater or lessor degree, now one and now another of these tendencies. On tho whole, capitalism is growing far more rapidly than before; but this growth is not only becoming more uneven in general, its unovenness also manifests itself, in particular, in tho decay of tho countries which are richest in capital.

The very substantial shifts in capitalism's material basis in its imperialist stage graphically confirm the fundamental character of these tenets of Marxist-Leninist theory, which help us better to understand the action of the long-term trends in the world economy that predetermine the features and course of its development in today's transitional period.

As the general crisis of capitalism deepens, the world capitalist economy more and more sharply experiences the effect of the socialist and national liberation revolutions, the mounting militarisation of the economy, and of numerous military shocks and upheavals. The development of 60 production in all the countries of this economy has been repeatedly interrupted or held back by economic crises, but the objectively operating long-term tendency toward growth of society's productive forces has still continued to make itself felt in the stage of imperialism as well. The economies of capitalist countries began to develop at particularly high rates after World War II, as witness the following facts. Fundamentally new branches of industry, transport, and communications were developed then under the impact of the scientific and technical revolution. Internationalisation of social production was noticeably broadened and deepened; the physical volume of international trade on the world capitalist market rose nearly eightfold.

The number of working people, and their vanguard the working class, increased on an unprecedented scale in the postwar decades. When Marx and Engels disclosed the very great revolutionary potential of the proletariat in the *Manifesto of the Communist Party*, there were only around nine million workers in the whole world. At the beginning of the 1980s their numbers had increased nearly 50-fold. The working class has become the leader of the broad masses' light for a revolutionary restructuring of society. And only in that struggle could the new social system arise that is going to replace capitalism.

Capitalism, Lenin said, creates its own grave-digger, itself creates the elements of a new system, yet, at the same time, without a 'leap' these individual elements change nothing in ths general state of affairs and do not affect the rule of capital.

These qualitative 'leaps' that deepen the crisis of capitalism were the socialist revolutions headed by the working class and the national liberation revolutions associated Non-socialist countries Years Socialist countries , total capitalist countries and their colonies *Area* 1900 1923 1981 16.0 26.2 100 84.0 73.8 89.0 05.0 23.5 *Population* 1900 1923 1981 7.7 33.0 100 92.3 07.0 70.0 60.0 17.5 *Sources*- statistical publications of the League of Nations and United Nations.

The patterns of the general crisis are not introduced into the world capitalist system from outside but are the natural result of its inner development and stem from the antagonistic nature of the capitalist mode of production itself, based on the exploitation of man by man and of some countries by others. The decisive consequence of the structural shifts under review was not a weakening but an unprecedented sharpening of the basic contradiction of capitalism, i.e. the contradiction between the constantly increasing social character of production and the capitalist, exploiter character of the appropriation of social labour. It is on that basis that all the other contradictions of modern capitalism and its world economy continue to develop.

A thorough study of the capitalist economy of his time enabled Karl Marx to conclude that the historical development of the antagonisms, immanent in a given form of production, is the only way in which that form of production can be dissolved and a new form established.

Lenin repeatedly stressed the paramount importance of extending this conclusion of Marx's to analysis of the monopoly stage of capitalism, and pointed out, already at the beginning of the contemporary transitional epoch that:

It would be impossible to put an end to the rule of capitalism if the whole course of economic development in the capitalist countries did not lead up to it.

Capitalist economics lias repeatedly tried in this connection to give battle to Marxist-Leninist political economy on the problems of world economic crises, in which, as Engels put it, 'the contradiction between socialised production and capitalist appropriation ends in a violent explosion'.

Many capitalist economists have tried with redoubled energy in the postwar period to demonstrate that capitalism would manage to enter a fundamentally new phase of development of the productive forces by means of new methods of economic policy and state-monopoly control, and that 63 the former instability and profound contradictions of the cyclic development of its economy would largely lose their force under the action of the capitalist state.

The very facts of capitalist statistics, however, prove that since World War II, and throughout the preceding history of monopoly capitalism, not a single capitalist country has been able to avoid considerable cyclic fluctuations and slumps in its economy. Periods of relatively high growth rates have inevitably alternated with periods of low ones, and sometimes even with an absolute fall in production and trade. The gap between the highest and lowest mean annual growth rates of the gross product was almost fivefold on the whole in the 50s and 60s, and in world industrial production and international trade it was even greater (from—2.5 per cent to -\-*Q.5* per cent and from—3 per cent to -\-*13* per cent respectively). The three economic slumps suffered by the world capitalist economy in the 70s and early 80s, including that of 1974–75, the deepest since the war, as was brought out iu detail at the 26th Congress of the CPSU, confirmed with all obviousness the irrefutable fact that state-monopoly control of the economy has proved incapable of curbing capitalism's elemental forces.

The new phenomena include an immense extension of the role of international monopolies in the economy of modern capitalism. It is their external economic expansion, unprecedented in scale and consequences, that lias largely determined the postwar features of the development of the world capitalist economy considered above. This expansion has acquired such a scale that the leading organs of the capitalist press and international economic organisations write about it with growing alarm. The special study *Multinational Corporations in World Development* made by UN experts said in particular:

In the past quarter of a century the world has witnessed the dramatic development of the multinational corporation into a major phenomenon in international economic relations.

The international monopolies, concentrating immense economic might in their hands, had gained control in the 70s of more than a liftli hotli of the production and the distribution of the aggregate social product of capitalist countries, and the lion's share of their direct foreign investments and half of their home and foreign trade. The Soviet economist T. Ya. Belous, basing herself on these facts, has remarked:

The international monopolies, which have developed as a result of the export and international interlocking of capital, are exerting an ever more decisive influence on the processes taking place in the economies and politics of modern capitalism. Their role in the postwar capitalist economy has increased steeply and continues to grow rapidly. No phenomenon of any significance in the economic and political affairs of capitalist countries, and in the whole system of their international relations, can now be fully evaluated and understood without analysing the activity of these monopolies.

Among the long-term factors determining the specific features of the development of the postwar world capitalist economy, as we have already remarked, a very material role is played by the scientific and technical revolution developing in the postwar period. Monopoly capital is endeavouring to put its main stake on it in the struggle to consolidate its class positions and to raise the effectiveness of the development rates of production. This revolution, by encouraging growth and internationalisation of capitalism's productive forces, has not simply led at the same time to reproduction on an ever greater scale and to an unprecedented sharpening of all former antagonisms immanent in the capitalist mode of production but has also given rise to new contradictions that are unresolvable for capitalism.

The new phenomena include (1) the ever-growing contradiction between the immense growth possibilities opened up 66 by the scientific and technical revolution and the obstacles that modern capitalism is putting in the way of their employment in the interests of society as a whole; (2) the bundle of contradictions between the social character of production, rapidly increasing under the impact of this revolution, on the one hand, and the state-monopoly forms of controlling technical progress, on the other hand; (3) the rapid deepening (along with further growth of the antagonism between labour and capital in the countries of imperialism) of the gulf between the interests of the overwhelming majority of the nation and those of the financial oligarchy. The unnatural position, in which production complexes often serving more than one country remain the private property of a handful of millionaires and billionaires, is particularly obvious in tin's situation. The need for the replacement of capitalist relations of production by socialist ones is becoming more and more pressing.

The scientific and technical revolution is also furthering the rise of new forms of the manifesting of the unevenness of world capitalism's growth. There has been a marked deepening of the gulf between various countries and groupings in the sphere of scientific research, and in the area of the application of the latest, most effective results of technical progress in the national economy. The U.S. monopolies broke away in this respect after World War II and forged ahead of the leading groups of the other capitalist powers. The 'technical gap' between the major and minor powers grew visibly, and also that between the imperialist countries as a whole and the developing countries.

By intensifying the spasmodic character of the economic growth of individual countries, the revolution at the same time inordinately heightened the imbalance of the various spheres of the capitalist economy. The contradictory character of the development of industry and agriculture, of the extraction and manufacturing industries, of the '65 traditional' industries and new ones, is growing. Under imperialism, when scientific and technical advances may be introduced mainly in the most highly monopolised enterprises, considerable changes arc taking place m the industrial structure of postwar capitalism. A number of the succeeding chapters will be devoted to analysing all these problems.

5

The Maturing of Capitalist Agriculture

We are all familiar with the classical story of how capitalism came to dominate industrial production and how capitalist relations of production swallowed up the individual artisanal producer. We recognize the power that the capitalist mode has to infiltrate and finally transform other forms of the organization of production and exchange. We sometimes think that the power of that transformation is so great that all of the significant action already occurred in the past, at least in Europe and North America, and was essentially over by the end of the nineteenth century. In the society we inhabit, it is a fait accompli, whose dynamic we can only understand by reconstructing the past because it is not happening around us. On second thought we realize that the transition was still in progress until very recently in a few skilled domains like medical care and entertainment, where individual artisans were able to ply their trade throughout most of this century, but these fossils of early capitalist relations seem exceptional because of their requirement of special talents or of necessary skills acquired by long training. But the view that the transition to mature capitalism is essentially over except at the margins of the main body of commodity production is clearly wrong, because it ignores an immense sector of basic essential commodity production, agriculture, which is still in the throes of the transition.

The penetration of capital into agriculture has been a long process of a different form than the classical case of industrial production usually exemplified by cloth weaving in the eighteenth and nineteenth centuries. Indeed, on the surface agriculture would seem to have been resistant to capital. After all, despite a 72 percent drop in the number of of individual farm enterprises in the United States from 6.7 million in 1930, there are still about 1.8 million independent farm producers today. This means that even though only 6 percent of these establishments account for 60 percent of the total value of farm production, there are over 100,000 separate enterprises producing more than half of all the value of the output. In the industrial manufacturing sector the four largest enterprises account for an average of 40 percent of value produced, and even in a highly differentiated product like clothing the top four companies produce over 15 percent of the value.

There has also been a major increase in the proportion of farmland that is leased to farmers who also own their own land. Roughly 55 percent of farmland is now operated by owner-renters who are for the most part small producers. Finally, despite the conventional wisdom that corporate farming is taking over, the proportion of farms and farmland operated by managers representing absentee owners has remained about 1 percent since the beginning of the century. Thus if we are to look for evidence of the capitalist transformation of agriculture, we will not find it in the classical industrial model. We do not find a concentration of more and more productive capacity in the hands of a very small number of farmers, employing a large wage labor force carrying out its tasks under close supervision and according to a tightly controlled schedule. There are, of course, some examples of a factory-like labor process in farming, especially in the harvesting of fruits and fresh vegetables, and these are often pointed to as evidence of a capitalist transformation to factory farming. However, the vast majority of farm enterprises do not employ a large labor force, but more typically have one or two hired laborers, usually for only part of the year.

In analyzing the process of the capitalist transformation of agriculture we must distinguish between farming and the agrifood system. Farming is the physical process of turning inputs like seed, feed, water, fertilizer, and pesticides into primary products like wheat, potatoes, and cattle on a specific site, the farm, using soil, labor, and machinery. The failure of classical capitalist concentration in farming arises from both financial and physical features of farm production. First, the ownership of farmland is unattractive to capital because it cannot be depreciated, and investment in farmland has very low liquidity as a consequence of the thin farm real estate market. Second, the labor process on very large farms is hard to control because farming operations are spatially extensive. Third, economies of scale are hard to achieve beyond what has already been realized by medium-scale enterprises. Fourth, risks from external natural events like weather, new diseases, and pests are hard to control. Finally, the cycle of reproduction of capital cannot be shortened because it is linked to an annual growth cycle in plants, or a fixed reproductive cycle in large animals. An important exception to this constraint has been in poultry, where there has been considerable success in shortening the reproductive cycle, and this has had important ramifications for the development of capitalist farming as we will see. For all of these reasons we do not expect to see, and have not seen, the wholesale direct takeover of farm ownership by large corporate enterprises employing large, well-controlled labor forces.

The agrifood system, however, is not simply farming. It includes the farm operation, but also the production, transportation, and marketing of the inputs to farming, as well as the transportation, processing, and marketing of the farm outputs. While farming is a physically essential step in the entire chain of agricultural production, the provision of farm inputs and the transformation of farm outputs into consumer commodities have come to dominate the economy of agriculture. Farming itself now accounts for only about 10 percent of the value added in the agrifood system, with 25 percent of the food dollar paying for farm inputs and the remaining 65 percent gained by transportation, processing, and marketing that converts farm products into consumer commodities. At the beginning of the century the value added on the farm was around 40 percent of the total food dollar, and many of the inputs were produced directly on the farm in the form of seed, draught animals, feed for the animals, manure and green manure for fertilizer, and family labor. For the most part these inputs are now purchased in the form of commercial seed, tractors, fuel, refined or synthesized chemical fertilizers, and machinery and manufactured chemical substitutes for labor. Thus, it is the production of farm inputs and the transformation of farm outputs that have provided an opportunity for industrial capital to capture profits in the agricultural sector.

Like any other industrial processes, the production of farm machinery, chemicals, and seeds, and the turning of threshed wheat into a box of breakfast cereal at the supermarket checkout counter are completely controlled by capital and its demands. The problem for capital, however, has been that sitting in the middle of the transformation of petroleum into potato chips is an essential step, farming, in the hands of two million petty producers. They cannot be dispensed with, they own certain essential means of production whose ownership cannot be concentrated (land in particular), and, while they are economically rational, they consume their surplus rather than turning it into capital. Agriculture is unique among all the sectors of capitalist production by possessing at its productive center an essential process organized around large numbers of independent petty producers. It is as if the spinning of yarn, the weaving of cloth, and the sewing of garments were in the hands of a few large capital enterprises (as they are), but the dyeing and finishing of the raw woven material were unavoidably the exclusive province of hundreds of thousands of home producers who bought the unfinished cloth and sold their product to clothing factories.

Farm producers have historically been in possession of two powers that stood in the way of the development of capital in agriculture. First, farmers could make choices about the physical process of farm production, including what was grown and how much, and what inputs were to be used. These choices, of course, were always constrained, partly because of local conditions of climate and soil, and partly because of the local nature of markets for farm products. Second, farmers were themselves traditionally potential competitors with the commercial providers of inputs, because they could choose to produce seed, traction power, and fertilizer themselves. The problem for industrial capital, then, has been to wrest control of the choices from the farmers, forcing them into a farming process that uses a package of inputs of maximum value to the producers of those inputs, and tailoring the nature of farm products to match the demands of a few major purchasers of farm outputs who have the power to determine the price paid. Whatever production risks remain are, of course, retained by the farmer. As the farmer loses any power to choose the actual nature and tempo of the production process in which he or she is engaged, while at the same time losing any ability to sell the product in an open market, the farmer becomes a mere operative in a determined chain whose product is alienated from the producer. That is, the farmer becomes proletarianized. It is of little import that the farmer retains legal title to the land and buildings and so, in some literal sense, is the owner of some of the means of production. There is no alternative economic use for these means. The essence of proletarianization is in the loss of control over one's labor process and the alienation of the product of that labor.

How has this transformation of farming been accomplished? In the first stages, in the century between the invention of reaping machines and the end of the Second World War, innovations in farming directly addressed that problem of the availability, cost, and control of farm labor through mechanization. No farmer could resist the arrival of the tractor, nor could one be home built. After the Second World War, refined and synthetic chemical treatments became the chief purchased inputs in the form of fertilizer, insecticides, and labor-saving herbicides. Again, these purchased inputs could not be resisted because of the large increase in yields and the reduction in labor. Herbicides, in particular, also reduced the requirement for tillage machinery, insecticides reduced the uncertainty of a successful crop, hormone sprays allowed for a close control of ripening time in fruit crops, and antibiotics prevented animal diseases. Once again there could be no competition between these industrial products and self-produced farm inputs.

The analysis of the growing role of capital inputs cannot be made, however, if a central feature of the productive process is lost sight of: the concrete use of all these inputs is to produce living organisms. The steps of mechanization and the use of chemicals were not possible in isolation from the nature of the organisms being produced. Unlike in other sectors of production, in agriculture living organisms are at the nexus of all input streams and are the primary sources of all output transformations. But living organisms are mortal, so their production requires their reproduction. That is, every cycle of farm production begins with seeds or immature animals to which value is added by on-the-farm operations, so seed (or the "seed" animal) is the central input into farming. The control of the biological nature of these seed organisms is a critical element in the control of the entire process of agricultural production, which puts the provider of this input in a unique position to valorize other inputs. For example, while a dramatic drop in the price of nitrogen fertilizer at the end of the Second World War made it economically possible for farmers to use this input in large quantifies, for this input to be useful it was necessary to breed plants, corn hybrids in particular, that could, in fact, turn a massive nitrogen application into crop yield. The successful mechanization of tomato harvesting was only possible by a close cooperation of the machine designers with plant breeders. The breeders completely remade the biology of the tomato plant, turning a loosely branched plant that flowers and sets easily bruised fruit continuously over the growing season into a short, stout, Christmas tree-like plant whose tough fruits all ripen at about the same time.

The consequence of the central position of seed input in the production process is that seed companies are potentially in an extraordinarily powerful position to appropriate a large fraction of the surplus in agriculture. There is a barrier to this realization however. The seed of a desirable variety, when planted by the farmer, produces plants that themselves produce yet more seed of the variety. Thus the seed company has provided the farmer with a free good, the genetic information contained in the seed, that is reproduced by the farmer over and over again in the very act of farming. Some way must be found to prevent the farmer from reproducing the seed for next year's crop. The historical answer to this problem was the development of the inbred/hybrid method of breeding, using hybrid crosses between inbred lines, which makes it possible to sell seed that will produce hybrid plants, but which themselves do not reproduce hybrids. Because the second generation would not be true hybrids and thus would lose yield and be more variable, the farmer must go back to the seed company every year to buy new seed. As a result of the immense profits made by seed companies selling hybrid maize seed, the method was spread into other organisms such as tomatoes and chickens. Moreover major commercial hybrid seed and chicken breeders like Dekalb, Funk, and Northrup-King were at one time acquired by pharmaceutical and chemical companies like Ciba-Geigy, Monsanto, and Dow, although subsequently there have been divestments and realignments. Only the largest hybrid seed company, Pioneer Hy-Bred, remained obdurately independent until, in 1997, 20 percent of its equity and two seats on its board were purchased by DuPont.

Generally, the ability of commercial seed companies to control seed inputs by the inbred/hybrid method was severely limited. First, the method cannot be made economically workable in many important crops like soybeans and wheat, or in large animals. Second, while the inbred/hybrid method was successful for general yield increase, large numbers of important specific characters such as resistance to particular diseases, or resistance to herbicides, or increases in oil content in oil seeds, do not show hybrid vigor, but must be introduced by other breeding methods. Third, there are characteristics that would be desirable to introduce into an agronomically important species, but which are present in other organisms that cannot breed with the species under cultivation. The most famous example was the desire to make corn plants able to fix nitrogen from the atmosphere, as legumes are able to do, by making their roots hospitable to nitrogen-fixing bacteria. While this would reduce the market for nitrogenous fertilizers, it would place the provision of nitrogen in the hands of seed companies!

The limitations on what changes could be made to agronomic species that would be profitable to seed companies and their chemical company partners or owners meant that the penetration of capital into agriculture had reached its apparent limits in the 1970s. The introduction of major new forms of mechanization into farm production had come to an end, partly because of the dramatic change in the cost of fuel and partly because a steady supply of immigrant labor that could be deported smiled progress in agricultural labor organizing. The growing public consciousness of the polluting effects of fertilizers and pesticides and the development of OSHA regulations to protect farm workers against the deleterious effects of insecticide and herbicide sprays discouraged radical changes in the uses of chemicals or even continued growth in the use of older materials. In addition, these fertilizers and pesticides were being used at very high rates, probably higher than could be economically justified by the farmer. There was, for example, no growth in fertilizers after 1975 or in synthetic pesticide application rates beginning in about 1980. Any further possibility for input providers and output purchasers to increase their appropriation of the surplus in agriculture depended on (1) making some radical changes in the biology of agronomic species, and (2) guaranteeing that such changed biological systems would remain within their ownership and control. Moreover, that appropriation could be greatly increased by a greater consolidation of both input and post-farm production sectors (purchasing, processing, and distribution), to provide near monopoly control. Enter biotechnology.

BIOTECHNOLOGY AND THE CONTROL OF PROPERLY

The claim is being made here that the purpose of the commercial use of biotechnology is to extend the control of capital over agricultural production. To accomplish this purpose biotechnological innovation should meet three criteria. First, the time and cost of its development must be within the limits set by capital investments in research. Thus, the attempt to introduce nitrogen fixation into non-leguminous plants has been largely abandoned by Agricetus, Agrigenetica, Biotechnica, and other biotechnology enterprises after spending over $75 million on the problem over more than ten years, despite the evidence that it ought to be possible, and despite the immense profits that could be made if it were successful. Second, the development must not provoke a significant challenge from politically effective forces concerned with health and environmental issues. All biotechnological innovations have been challenged on the basis of environmental and health risks, and this contributed significantly to the demise of at least one early biotech project. An important impetus for introducing biotechnology is that the resistance to yet further applications of fertilizer and pesticides was impeding further increase in the appropriation of the surplus in agriculture by input producers. Third, ownership and control of the product of biotechnology must not pass into the hands of the farmer but must remain with the commercial provider of the input.

The requirement that the biotechnological innovator maintain ownership and control over the altered variety creates a contradiction. As previously discussed, the farmer acquires a free good, the genetic information contained in the seed, when he or she purchases a new variety, and the breeder loses its ownership. The property rights protection offered by the inbred/hybrid method is limited to a few organisms and a few agronomic characteristics, and biotechnology has been introduced in precisely those instances where the inbred/hybrid method does not apply. How, then, can breeders appropriate a greater share of the surplus when they are giving away the critical material, the genes? The answer has been provided by a combination of legal and biological weapons in the hands of the breeders. These weapons are legal rights granted to breeders by the Plant Variety Protection Act and subsequent court decisions, in combination with the use of standard DNA "fingerprinting" that allows an unambiguous determination of the source of farm products. It is now standard that a farmer who wishes to purchase a bioengineered seed must sign a contract with the seed producer giving away all property rights in the next generation of seed produced by the crop. Not only does the farmer undertake not to sell seed from the crop to other farmers ("brown bagging"), but more revolutionary, the farmer is prohibited from using the next generation of seed to produce next year's crop on his or her own farm. All farmers who buy seed of Monsanto's Roundup Ready soybeans, or that company's seed potatoes for a special variety that makes "light" potato chips with low oil retention, must, by the terms of the contract, return to Monsanto in the next season if they wish to continue production of those varieties. (Monsanto is the producer of Roundup, a potent herbicide that kills all plants including soybeans. "Roundup Ready" soybeans, produced by genetic engineering, can be grown in fields heavily treated with Roundup without killing them and, presumably, without materially affecting their yield.) The enforcement of such a contract depends on the ability of Monsanto to identify a crop, and this can easily be done from a single plant or even a single seed because the DNA of the engineered variety contains certain characteristic sequences, placed there deliberately by the genetic engineers, that are unique to the variety. The assay of crops for such labeled sequences is called "genome control" by the biotechnology laboratories of seed producers, and a considerable laboratory effort has been put into developing these detection techniques. Nevertheless some brown bagging and replanting has been taking place. In reaction, Monsanto has placed full-page advertisements in magazines read by farmers, threatening and cajoling:

> When a farmer saves and replants Monsanto patented biotech seed, he understands that what he is doing is wrong. And that, even if he did not sign an agreement at the time he acquired the seed [that is, replanted or bought "brown bag" seed from a neighbor], he is committing an act of piracy.... Furthermore, seed piracy could cost a farmer hundreds of dollars per acre in cash settlements and legal fees, plus multiple years of on-farm and business records inspection.

It only takes a few widely publicized legal judgments to keep the rest in line.

But the story of property rights has yet one more chapter. The inbred/hybrid method only applies to a few organisms, and the contract system requires threats, monitoring, and litigation to make it work. It is biotechnology that has now perfected the solution to ownership in seed crops. It was announced on March 3, 1998 that a patent had been granted for a genetic manipulation that would allow plants to set seed and therefore make a crop, but which would render those seeds unable to germinate. Thus, at one blow, the problem of capitalist seed production, first addressed by the invention of the inbred/hybrid method at the beginning of this century, has been solved for all seed crops. As the inventors point out, there is still development to be carried out before this bit of biotechnology becomes a commercial reality, but there seems no bar to its transfer to any crop. And who are the inventors and owners of this patent? They are the Delta and Pine Land Company, a leading breeder and producer of cotton seed and soybean seed, and the Agricultural Research Service of the United States Department of Agriculture. Yet there is no suggestion that this development will be of any benefit to farmers or consumers. We could hardly ask for a more blatant case of state support of private property interests to the exclusion of any public benefit.

The use of the contract to enforce breeders' property rights allows us to make some predictions about the limitations of genetic engineering. At the present time the hormone BST, which causes dairy cows to direct more of their metabolism to milk production, is produced commercially by Monsanto in fermenters using genetically altered bacteria. But cattle normally produce their own BST, and there is no reason that the regulatory DNA that controls production of this protein in cows could not be altered to increase the amount. This would then make the purchase and administration of commercial BGH unnecessary. We can predict, however, that this is unlikely. First, dairy herds have always been largely self-reproducing on small- to medium-sized enterprises, and there are no major commercial dairy herd breeders equivalent to major seed companies. Second, enforcement would be very difficult. It is easy for a representative of Monsanto to "acquire" a single potato or a few seeds from any farmer's field, or from a local elevator. It is considerably more intrusive to take the blood or tissue sample from a farmer's dairy herd that would be needed for "genome control." Moreover, since dairy herds are not all reproduced at one time but have overlapping generations, it would be impossible to say, except after a number of years, whether a cow was one of the originally purchased stock, or an offspring.

PRODUCTION CONTRACTS, BIOTECHNOLOGY, AND THE CONTROL OF FARMING

If the only effect of biotechnology and the contract system of guaranteeing property rights were to extend the domain of manufactured inputs into farming, nothing very revolutionary would have occurred. Farmers for a long time have been the purchasers of manufactured inputs. The major structural changes that are occurring in agriculture arise from a vertical integration of farm production in such a way that the purchasers of farm outputs take control of the entire production process. This vertical integration is made possible by 1) a technical linking of the inputs and outputs, 2) the dual function of a single capital enterprise as both the monopsonist (near monopoly) purchaser of outputs and the provider of critical inputs, and 3) a contract mechanism that links farmer into the loop of inputs and outputs. The use of such contracts predates biotechnology. Wherever the purchaser of farm outputs is also the processor of those outputs for the market, the possibility of vertical integration has existed. Contract farming has been a common feature of vegetable production for canning. Tomato canneries in Ohio were built in a location central to the farms, the canning company provided the seed and chemical inputs and collected the ripe tomatoes. The farmer provided the land and labor. But the system has evolved greatly since the first canning contracts. The critical role played by biotechnology has been in the material linking of inputs and outputs. In order to guarantee an efficient integrated system of production, the biological inputs into the chain of production, the organisms being raised, are engineered to fit the package of other inputs, the mechanics of the farming process, and the qualifies that the final output is to have for the market. While some of these aims can be accomplished by conventional methods of breeding organisms, many of the needed qualifies, such as specific disease resistance or qualitative changes in the composition of the organism, are best produced by biotechnological manipulations. Moreover various cloning and cell culture techniques make it possible to reduplicate large numbers of input organisms with desired heritable qualifies, no matter how those qualities were originally produced.

An example of the nature of contract farming is in the production of broilers (chickens raised as meat) where the system is especially entrenched. A major supplier of chickens to supermarkets and fast food restaurants is Tyson Farms of South Carolina. Tyson chickens are produced, not by Tyson "Farms," but by small farmers, owning about 100 acres, producing an average of 250,000 chickens per year, with a gross income of about $65,000 and a net of around $12,000.

This production is under a four-year contract with Tyson (or other similar regional firms), a contract that makes Tyson the sole provider of the chicks to be raised, the feed, and veterinary services. The company is also the sole determiner of the number, frequency, and type of chicks provided. Tyson then collects the mature birds after seven weeks, at a date and time of their own determination, providing the scales on which the birds are weighed and the trucks to take them away. The farmer provides the labor, the buildings in which the chicks are raised, and the land on which the buildings stand. The detailed control of inputs and farming practices are entirely in the hands of Tyson. So, "The Producer (farmer) warrants that he will not use or allowed to be used ... any feed, medication, herbicides, pesticides, rodenticides, insecticides or any other items except as supplied or approved in writing by the Company." Moreover, the farmer must adhere to the Company's "Broiler Growing Guide" and a failure to do so puts the farmer into "Intensified Management" status under the direct supervision of the Company's "Broiler Management and Technical Advisor."

The chicken farmer has ceased to be an independent artisan, buying materials, transforming them by his or her labor, and selling the product on a market. The contract farmer buys nothing, sells nothing, nor makes any decisions about the physical process of transformation. The farmer does own some of the means of production, land and buildings, but has no control over the labor process or over the alienated product. The farmer has then become the typical "putting out" worker characteristic of the first stages of capitalist production in the seventeenth and eighteenth centuries. What the farmer has gained is a more stable source of income, at the price of becoming an operative in an assembly line. The change in the farmer's position from an independent producer, selling in a market with many buyers, into a proletarian without options, is reflected in the nature of the recommendation in the 1998 report of the National Commission on Small Farms, that Congress should amend the AFPA [Agricultural Fair Practices Act] to provide the USDA with administrative enforcement and civil penalty authority that will, in turn, enable growers to to organize associations and bargain collectively without fear of discrimination or reprisal. (emphasis added)

The combination of biotechnological manipulation and contract farming also can have a catastrophic effect on third world economies. Much of the import of agricultural products from the Third World consists in qualitatively unique materials like coffee, flavorings, essences, and food oils with special properties. Moreover, the production of these materials is at a low technological level with high labor inputs, in countries with unstable political and economic regimes. As a result, the price and availability of, say, palm oil from the Philippines are unstable. These characteristics make such agricultural products prime targets for gene transfer into domestic species which will then be grown as specialty crops under contract to processors. Calgene has engineered a high lauric acid canola (rape seed) strain for oils that are used for soaps, shampoos, cosmetics and food products that formerly required imported palm oils. These special canola strains are now produced in the Midwest under contract, displacing Philippine production on which a large fraction of the rural population depends economically. And the genes for the biosynthesis of caffeine have been successfully transferred to soybeans. If the essential oil genes for coffee flavor can also be transferred, then Central and South America and Africa will lose their market for beans destined for powdered coffee.

It would be a mistake to think that agriculture has followed the classical picture of the spread of capitalism. Unlike in industrial production, the first step in the capture of agriculture by capital was the immense flowering of input industries and output processors, who appropriated the surplus in agriculture by selling the petty entrepreneurial farmer what he needed, and buying what he produced. No parallel exists in the industrial sphere. It is only with the saturation of that possibility of appropriation that wholly new techniques have come into play. By concentrating on the central material link in farm production, the living organism, which at the same time was the most resistant to capitalization, biotechnology has accomplished two steps in the penetration of capital. First it widened the sphere of input commodity production by including a wide array of organisms that had previously escaped. Second, and more profound, it is making vertical integration possible with the accompanying proletarianization of the farmer. It is this second stage that is the capitalist agriculture of the future, because the physical nature of farm production, inevitably tied to the land, is such as to maintain its unique organization as a productive process.

AGRICULTURE AND MONOPOLY CAPITAL

The Evolving Food System of the United States

Native Americans and the Europeans settlers who subsequently occupied the territory of the United States developed an agricultural and food production system that was largely self-sufficient. Most families produced, processed, and consumed their own food. Families made many of the tools and produced most of the seed they needed, and raised their own animal power. Few items were purchased for food production and processing, but there was very little surplus food or fiber to sell. The family controlled its food system from seed to plate - the ultimate integrated food system. The purpose of colonies, however, was to send raw materials including food and fiber products back to the mother country. The industrial revolution and the development of industrial cities, first in England and then in the United States, required that farmers produce a larger and larger surplus of food for the growing urban market. Government policy encouraged farmers to produce an ever greater excess of food and fiber and to do so with less and less labor. Thus agriculture evolved from a subsistence agriculture to a commercial agriculture in which the role of the farm family was to produce for the market.

In the evolving commercial agriculture system, farmers became separated from the consumers. Increasingly, farmers sold their products to firms which would process and transport the food and fiber to the distant populations. These firms were eager to serve the function of linking the farmer to the consumer. Although relatively simple at the outset, the food system soon began to develop into many components, or stages, as a result of the specialization of function that is a characteristic of the industrialization model. As specialization evolved, the early "seed to plate" integrated food system developed into a multi-stage food system with hundreds of firms competing at the various stages. As the system developed more stages, the middlemen, as they were often called in the farmers' movement literature, were not just the firms between the farmer and the public, but also included those enterprises that provided a growing number of inputs such as credit and farm equipment. As the food and fiber products moved from one stage to the next within the processing-distribution sectors, information was available on the price of the product. Many farm products were auctioned at public events where anyone could observe or bid. This was the agriculture and food system which was often held as a model of a competitive system in which 1) no firm bought or sold enough of the total goods or services to influence the price; 2) there was relatively easy entry and exit from any stage; and 3) information regarding the price of the goods and services along the total food chain was available to all. This food system in which the family farm structure was an important component was a good example of the early stage of capitalism. Later, the two processes of horizontal and vertical integration would begin altering the power relations between firms. This led to increased concentration of capital resources, and with it, increased concentration of control in the food system and the decline of the family farm structure.

Capital Concentration and Large Commercial Agriculture

Although elements of the above ideal type of a competitive food system could be found in parts of the United States at various times, it never did exist in many areas of the country. For an individual farmer, the question was not how many firms were involved in the different stages of the food chain across the country, but rather how much of this commodity chain was accessible to his (sometimes her) farm. AS farmers moved west, one of their major problems was how to transport their products to markets in the eastern cities. The government, wishing to promote increased industrialization, also perceived the problem and subsidized the construction of transportation systems, especially railroads. This often made the farmer dependent on a monopoly which could exploit him/her because of the unequal balance of economic power. If a farmer had access to only one railroad, the power relationship certainly favored the railroad. That farmer faced a monopoly regardless of how many other railroads existed in the country. Thus, many farmers faced the issue of monopoly control of capital from the time they became commercial farmers and began to be dependent on a single transportation system to move their products to the market.

Railroads in some parts of the country needed the business of farmers, but they had access to hundreds, if not thousands, of farmers. They were not dependent on any single farmer. Their only concern was that farmers might be successful in organizing a united stand against the railroads. Frank Norris' agrarian populist novel, The Octopus: A History of California (1901), was about a handful of farmers who attempted such a united stand, rising up in direct rebellion, only to be dispatched by the railroads. The whole history of the farmers' movement is largely about the unequal power balance between farmers and the railroads, and, more generally, between the farmers and all the "middlemen" they depended on for transportation, markets, and a host of inputs such as credit and farm equipment.

In southwestern Minnesota, for example, the selling of water fowl to cities to the East and later the selling of agriculture products, especially grain, all depended on the railroad for transportation. In the case of grain, the farmers also depended on the elevators (large silos) to store and to transfer it from farmers' wagons to the train. Many of the transnational corporations (TNCs) of today, like Cargill, exercised economic power in many of the local markets in which they began operating. Most farmers' movements were not successful in establishing alternative economic systems or firms which benefited the farmers, but with the help of legislation to encourage farmer cooperatives, there were some successes.

Horizontal Integration

Most food firms started as relatively small, local firms, but as they became profitable they expanded their operations into other geographic areas. The expansions occurred through building new facilities, acquisitions, and mergers. The expansion of a firm within the same stage of the food system as their original operation is called horizontal integration. For example, the increase in size and decrease in the number of farms in the United States during most of this century is an example of horizontal integration. Horizontal integration also occurs at each of the other stages of distribution and processing. Although there are great variations among the different commodity sectors regarding the ways concentration of ownership and control have occurred in processing and distribution stages, the same general pattern of fewer and larger firms in each stage has been underway during the last half of the twentieth century in the United States and has become most obvious during the last decade at the global level.

In some commodity sectors, one can point to significant concentration of the processing firms in even the first half of the century. For example, pork and beef slaughtering and processing were dominated by Wilson, Armour, and Swift as we entered the twentieth century. Opposition to their practices in the Chicago stockyards inspired Upton Sinclair's The Jungle (1906) and led to the passage of the first Food and Drug Act that same year. And their collusion to set monopoly prices was largely responsible for the creation of the Packers and Stockyards Agency of the U.S. Department of Agriculture (USDA) in 1921 to monitor predatory practices. The Swift and Armour brand names exist today, but the firms were bought by ConAgra, which also bought Miller and Monfort. Some would argue that the fact that these firms do not exist today suggests that even firms with significant economic power can themselves be eliminated. But the important point is that they were absorbed within the larger agglomeration of capital. Still it is true that with the continuing trend toward concentration and centralization of capital no firm is safe from takeover or elimination in other ways. (This is often used as justification for why no government action should be taken to interfere with the free market.)

Forty percent or more of the processing of all agricultural commodities in the Midwest are controlled by the four largest firms. Although debate continues in the United States and in other countries on what constitutes an oligopolistic or near oligopolistic market, much of the economic literature suggests that when four firms control 40 percent of the market, they are able to exert influence on the market unlike that in a competitive system. In the meat sectors 87 percent of the beef cattle are slaughtered by the four largest firms (81 percent by the largest three) and 73 percent of the sheep are processed by the four largest firms. The control of hog slaughtering by the four largest firms increased from 37 percent in 1987 to 60 percent today. Over one half (55 percent) of the broilers (chickens produced for meat) today are produced and processed by the four largest firms, with Tyson now producing and processing almost one third of the broilers in the United States. In the crop sectors, the four largest firms process from 57 percent to 76 percent of the corn, wheat, and soybeans in the United States.

Like the narrow opening of an hour glass which controls the flow of sand from the top to the bottom, the processing firms are positioned between the thousands of producers and millions of consumers in the United States and the world. These firms have a disproportionate amount of influence on the quality, quantity, type, location of production, and price of the product at the production stage and throughout the entire food system. The only stage in which a set of firms begins to equal the economic power of the food processors is the retail stage, which is also becoming more horizontally integrated. The interface between the processing and retail stages is currently where the giants of the food system interact. Certainly, it is not an area characterized by easy entry and exit. How many firms in the world have sufficient capital to face the economic power of these two sets of firms?

Advantages of Cross-subsidization to Concentrated Capital

In the food system, horizontal integration usually refers to expansion in the same stage of the same commodity sector. However, if one considers the sector to be meat, then horizontal integration would include the total meat sector. For example, ConAgra ranks in the top four firms in the processing of beef, pork, broilers, sheep, turkeys, and seafood (which is not on the list). Spokespersons for the industry frequently highlight the competition for the public's dollar between different meats, such as the competition between beef and poultry. They frequently use the competition between meats to argue that the producers must make certain changes in their practices. This competition between commodities is also frequently used to justify the check-off system in which a per-animal fee paid by farmers is used primarily to support product promotion and research. The cost of the check-off-system is borne by the producers, but there is growing concern as to who is benefiting. Currently the Livestock Marketing Association is leading a petition drive to force a beef producer recall referendum on the check-off programs. Their argument is that "after $1 billion spent in promotion and research over the first 10 years of the program, beef demand is still declining" (Feedstuffs, March 2, 1998, p.5). They suggest the check-off funds are being spent directly or indirectly on projects that benefit processors and retailers rather than beef producers, and that the results are increased concentration and integration of the industry. One can ask how much competition exists between the different meat products when key decision-makers are involved in more than one part of the meat sector.

The movement of firms into the processing of several commodities may at one level be an extension of horizontal integration, but it also represents a major qualitative change in the economic power relationships. When a firm has a dominant position in several commodity systems, it can cross-subsidize. Firms operating in more than one commodity system gain economic power because they can survive a major loss in one commodity system over a long period of time if they are making significant profits in other systems. If a loss continues very long in a single-product firm's only commodity, it faces serious financial difficulty.

Lane Poultry was the largest broiler producer and processor in the United States following its purchase of Valmac Industries in 1980, but it was still a single-product producer. Lane lost Valmac Industries and then was itself purchased by Tyson Foods because of its economic losses in nine of eleven consecutive quarters in the late 1970s and early 1980s. Being the largest firm in a commodity sector, but a single product producer, does not assure enough economic power to survive. Larger firms with profits in other sectors or systems have more economic power and may overtake them. Information we obtained from executives of a couple TNCs involved in broiler production at the time indicated that the goal of their firms was to obtain a larger share of a growing market. (This was a time when per capita consumption of poultry was increasing rapidly while beef consumption was decreasing in the United States)

Planned overproduction and selling below cost of production also occurred in the farm-raised catfish sector early in the 1980s. Two major catfish cooperatives, Southern Pride and Delta Pride, experienced the problem of competing against ConAgra, Cargill, and Chiquita, the parent company of Morrill (now owned by Smithfield). The three TNCs were able to cross-subsidize. The cooperatives survived despite the fact that the annual report of one of the TNCs showed a loss in the catfish division for two years because of "overproduction" in the sector. Conversations with some of the TNC's personnel indicated the firm was prepared to extend this loss for another year or two. We concluded that because the five firms absolutely dominated the production and processing of farm-raised catfish at the time, the low prices were the result of overproduction and an effort on the part of the TNCs to gain market share at the expense of the members of the cooperatives. In fact, Cargill, which has now exited the sector, entered the catfish sector during the time of negative profits with plans to increase catfish production. Like the broiler and catfish sectors in the past, the hog sector is currently involved in a large increase in production even when prices are low and are predicted to stay low. The issue is market share, not efficiency. Large firms that can cross-subsidize can operate in this arena, but smaller, nondiversified firms cannot survive. Economic power, not efficiency, predicts survival.

Vertical Integration

The second major strategy of monopoly capital is vertical integration. Vertical integration occurs when a firm increases ownership and control of a number of stages in a commodity system. Just like diversifying into different commodity sectors (horizontal integration), this structure gives the firm more economic power.

Another example of the extent of vertical integration in the food system comes from ConAgra's annual report. ConAgra indicates that it is the largest distributor of agricultural chemicals in North America, one of the largest fertilizer producers, and in 1990 it entered the seed business. (Since then it has formed a joint venture with DuPont and formal relationships with some of the seed companies involved in biotechnology). ConAgra owns 100 grain storage elevators, 2,000 railroad cars, and 1,100 barges. ConAgra is the largest turkey producer and second largest broiler producer. It produces its own poultry feed, as well as other livestock feed. It also owns and operates hatcheries. ConAgra hires growers to raise its birds and then it processes the birds in its own facilities. This broiler meat can then be purchased as fryers under the name of Country Pride or in further processed foods such as TV dinners and pot pies under the labels of Banquet and Beatrice Food. From the basic raw materials for agricultural production to the retail store, a significant proportion of the food system is owned and controlled by ConAgra. ConAgra is the second largest food firm in the United States (behind Philip Morris) and the fourth largest in the world, with operations in thirty-two countries.

In the subsistence food system, the family controlled its food from seed to plate. In the emerging vertically integrated food system, a few food companies are gaining control of the country's food system by controlling it from seed to shelf. This system is being extended around the world by many of the firms that are headquartered in the United States.

Starting in the 1950s and 1960s, when feed companies and others started hatching their own baby chicks, hiring growers to provide labor, buildings, and land, and constructing their own processing facilities for broilers, the farm press and farm community began to focus on contract production. Contract production is very different from forward contracting of a commodity product. Forward contracting is a sales agreement between a farmer and a buyer that involves an agreed upon price and other terms of the sale to be carried out at some future date. Contract production is an industrial model in which the integrating firm outsources a needed ingredient - the agricultural raw product.

In contract production, the growers are required to provide the land and the buildings, and equip the buildings to the integrating firm's specifications while providing all of the labor for the production stage of the system. The growers are thus hired workers paid on a piece rate basis. They never own the birds or the feed, and have no knowledge of the genetics or the feed ration. The integrating firms provided the birds, feed, and medication. All of the major decisions are made by the integrating firm. The growers mortgage their land to raise the capital to build the buildings, which cost over $100,000 each. Typically their repayment schedule extends over a ten- to fifteen-year period, while the contract with the integrating firm goes from one batch of chickens to the next - a period of about six weeks. By the time the buildings are almost paid off, the equipment needs to be replaced and the buildings need to be modernized. As a consequence, few growers ever get out of debt. It is estimated that although about one half of the capital in the broiler sector comes from the growers, all of the major decisions are made by the integrating firms. The growers are well aware that they can be cut off at any time.

In the early stages of vertical integration in the broiler industry, most growers had access to several integrating firms, but over time the numbers were reduced. For example, in 1969 in Union Parish, Louisiana, there were four integrating firms. Two were locally owned feed operations, and two were operations based out of state. By 1982, the two local firms were no longer integrating firms; they were now growers. The two outside firms were owned by ConAgra and Imperial Foods, one of the largest food companies of England. Within the following year, ConAgra bought the Country Pride broiler facilities from Imperial Foods. The growers report that they have had no price increase since 1982.

Two processes occur that alter the growers' opportunities. First, as the integrating firms in a given geographic area become fewer, the power relationships between growers and the firms become more unequal. Because of transportation and other costs, most integrating firms will send trucks out only about 25 to 30 miles from the processing site to deliver feed and to pick up poultry for processing. Today there are about 40 firms producing about 97 percent of the broilers in this country. In total, they operate about 250 processing facilities. Thus, there are very few growers who live close enough to more than one processing facility to even have an option to choose between integrating firms.

The second process that limits options for growers is that as the number of firms operating in the same geographic area declines to two or three, an informal agreement evolves between integrating firms that they will not raid their competitors' growers. If a grower gets cut off from one integrating firm, they cannot enter into a contract with another. My thirty years of observing the poultry sector suggest that in early capitalism when the growers have access to several integrating firms, the growers experience financial success, but when the system moves to monopoly capital the growers find themselves in financial crisis. The courts have also found that the growers are at the mercy of the integrating firms in other ways. Errors in the weighing of both feed and poultry have become so well documented that legislation has recently been introduced to address this issue. The USDA has also made a commitment to study this problem.

Globalization and Monopoly Capital

A third way TNCs can cross-subside and gain economic power is by operating in many different countries of the world. A firm like Cargill, which has operations in seventy countries, can absorb a loss in one or two countries over a few years as long they have good rates of return in other countries. It is easy to see how TNCs can gain footholds in new countries. They begin by producing or processing a commodity and thus increase the supply of the product which, assuming a given demand, drives the price down. In so doing, the price can drop below the cost of production. A local firm, especially if it is a single product producer, can't absorb such a loss for long, but a TNC can absorb the loss for a relatively long time.

The food systems of the world are becoming so integrated by the TNCs that it makes little sense to speak of a food system in a single country. For example, not only do IBP, Cargill, and ConAgra process 81 percent of the beef in the United States, they also now have feedlots and slaughtering facilities in Canada and about the same market dominance there. With the passage of the North American Free Trade Agreement, beef can easily travel back and forth across the border. It is possible for a 450-pound calf to be purchased by one of the three firms on one side of the border and sent across the border to be grown to 800-pounds in a contract arrangement with a rancher. The firm may then decide to move the animal across the border again to be finished in their feedlot. If the firm does not have sufficient capacity in their slaughtering facility in that country, it may move the animal across the border again to be slaughtered. The meat may once again be moved across the border to be consumed. The question then is, in what country was the animal produced?

These same firms have beef operations in many other countries, including Australia, Mexico, Brazil, and Argentina.

In such a system, the opportunity for governments to control their food system is diminished and the role of TNCs is increased. As global trade barriers are reduced, it becomes increasingly difficult for relatively small local firms to compete with the economic power of the TNCs. Maple Leaf Foods, Canada's largest food processor, is currently struggling to survive in the pork sector. In order to survive, Maple Leaf has lowered wages, leading to labor strikes, in an effort to compete with the low wages paid by the U.S. slaughtering plants that have been obtaining live hogs from Canada (Feedstuffs 3/23/98). Smithfield, the largest hog slaughterer in the United States, has indicated an interest in buying the Maple Leaf facility in Ontario, one of the facilities with major labor discord.

Many TNCs involved in the global agriculture/food system are conglomerates. Cargill, which is such an important player in meat and grain processing, also makes barges, operates barge and shipping lines, and processes pig iron, scrap iron, and other metals and petroleum products. Philip Morris, which most people associate with tobacco and cigarettes, is the largest food corporation in the United States and second largest in the world. It sells Kraft General Foods products (including Kraft cheese), Maxwell House coffee, Miller Beer, and Oscar Meyer meats. In 1996, Philip Morris had over $16 billion in revenues within the United States from its North American Food Division (Richmond Times-Dispatch, April, 19, 1997). Many of the conglomerates regularly buy and sell various divisions. Purina Mills, the largest animal feed firm in the United States (with 10 percent of the market), has changed hands several times. Wishing to focus on its pet food segment, Ralston Purina (now the world's largest manufacturer of dry dog and cat food and of batteries and flashlights), sold its Purina Mills Division in 1986 to British Nutrition, the largest livestock feed producer in Europe, which is owned by British Petroleum. In 1993 British Petroleum sold Purina Mills to the Sterling Group of the United States. Then in early 1998, the Sterling Group sold Purina Mills to Koch Agriculture Company, a subsidiary of Koch Industries. Koch Industries is a diversified energy and industrial firm ranked the second largest privately held company in the United States. Vertical integration, horizonal integration, integration among various segments of the food systems, conglomeration, and global integration all lead to increased economic power for the TNCs.

Community Consequences of Absentee Corporate Ownership

Capital that comes from outside the local community has major economic consequences for the local community. In small, locally owned businesses, such as family farms, family clothing stores, and family grocery stores, the family subtracts its annual expenses from its income to determine profits that are then allocated among labor, management, and capital. For the economic well-being of the family and the rural community, it makes little difference how the profits are allocated among the three costs of production, because the family spends much of the profit in the local community.

In the past, when family businesses were the predominant systems in rural agriculture communities, researchers talked about "multiplier effects" of three and four. Money generated in the agricultural sector would circulate in the community, changing hands from one entrepreneurial family to another three or four times before leaving the rural community. This greatly enhanced the community's economic viability.

Giant corporations headquartered in distant places see labor as just another input cost to be purchased as cheaply as possible. Their profits are usually immediately taken out of the local rural community. They go to the firms' headquarters and on to stockholders or, if the corporation is a TNC, the profits are very likely invested in the food system somewhere else in the world.

Only the return to labor, which was bought as cheaply as possible, remains in the rural community. Rural mining towns in Appalachia and the West show the adverse consequences of absentee ownership under a regime of concentrated capital.

Today, the economic impacts of agricultural production on communities are perceived to be so small that few rural economic development specialists see any hope in expanding the economic base of a rural community by focusing on the production stage of the food sector. In some areas, attention is focused on the processing sector that is referred to as "value added."

These economic outcomes have major social consequences for communities, especially when combined with the social consequences coming from changing the workers' relationship to their work setting.

Fifty years ago, Walter Goldschmidt's famous study in California showed a strong relationship between the structure of the food system and the social condition of the community, revealing that the well-being of communities dominated by large-scale, abstentee-owned, corporate farms was greatly inferior to that of communities in which family farms predominated.

A host of studies building on Goldschmidt's work have confirmed these results. But the social costs, usually called externalities, are of no concern to TNCs, which see increasing the wealth of their stockholders as their major goal.

Perhaps the ultimate political power these giant food firms have is that - like some saving and loan associations, banks, automobile and airplane manufacturers - they are perceived to be so vital to the country that their bankruptcy would lead to major social disorganization. Given that food is a necessity, is there any doubt that the federal government would provide the resources to keep these firms from collapsing?

Monopoly capital does not behave differently in the food system than in any other segment of the economy. Agriculture is unique primarily because it took monopoly capital so long to dominate the sector.

The fact that agriculture is losing its uniqueness is perhaps best revealed by the fact that Mitsubishi, among the largest automakers and the second largest bank in the world, is now one of the world's largest beef processors.

The "mainstreaming" of the food system into the larger capitalist system is also revealed by the fact that Pioneer Hi-Bred, DeKalb, Mycogen, and other seed companies which own the property rights to the new varieties created through the use of biotechnology are now forming new organizational relationships with chemical firms (e.g., Novartis, Monsanto, DuPont, Dow) and with firms which ultimately process the genetically modified new varieties (e.g., ConAgra, Cargill). One can predict that up against such power, the yeoman crop farmer will soon resemble the broiler grower.

To give an idea of how rapidly the situation is changing, during the month following completion of this manuscript, Monsanto announced the purchase of Delta and Pines Land Company and DeKalb seed companies and now controls 85 percent of the cottonseed sales as well as the patent for the terminator gene (that allows a company to sell seeds that farmers can plant and get a crop, but seed saved from the crop will not grow the following season). Further, Monsanto developed a joint venture with Cargill to further integrate the input and processing stages for grains and then announced a merger with American Home Products to form a $96 billion firm with expected sales of $23 billion in 1998.

6

Aspects of the Agrarian Situation and Alternative Policies

GROWING CRISIS IN AGRICULTURE

An unprecedented and all-encompassing agrarian crisis is deepening. India is moving fast towards severe food shortage and starvation. The per capita availability of food grain has declined and has reached a level unprecedented in the last five decades. The growth rate of productivity and production in agriculture is also declining. Unemployment is growing fast. The number of days of work available to a worker in a year and the wages of agricultural workers are declining. Indebtedness is growing. Many sections of the rural poor are being marginalised. The increasing migration of agricultural labourers, poor and middle peasants from the rural areas is causing serious social and economic problems. The steep fall in prices of agricultural commodities and the increase in prices of agricultural inputs have adversely affected the interests of all sections among the peasantry, particularly the poor. Poverty is spreading to new sections and areas. Starvation deaths and peasant suicides are increasing. Unevenness in growth and development, sectoral and spatial, is increasing.

Since independence, India has been able to make substantial gains in agricultural production, including cereal production. This has been made possible because of increases in aggregate area sown, the expansion of irrigation facilities, certain land legislation, the introduction of high-yielding seed, the use of fertilisers, improved implements, farm machines, and technology for pest management, a price policy based on minimum support price and procurement operations, infrastructure for storage and cold storage, increase in investments, improvements in the trade system and other factors. These achievements, however, have not stopped the growing crisis in agriculture; the crisis is caused by inherent structural defects in production relations, and the weakness of the path of development chosen as well as certain specific measures taken in the agrarian sector.

Land Reforms

Land legislation passed and implemented in States by bourgeois-landlord State Governments have failed to achieve genuine land reform. Left-led State Governments in West Bengal, Kerala and Tripura alone have made substantial changes in land relations in the interests of the overwhelming majority among the peasantry. In other States, land concentration remains more or less unchanged. Breaking land monopoly and distributing land gratis to agricultural labourers and poor peasants, destroying the structures of traditional systems that hamper the process of development, such as caste and gender oppression and servitude and bondage are prerequisites for releasing the creative energy and labour enthusiasm of millions of peasants. Such reform also helps the political need to establish economic and political citizenship for the excluded masses in order to ensure their integration into a cohesive nation.

The land legislation of bourgeois-landlord governments did not aim to abolish landlordism but to transform semi-feudal landlords into capitalist landlords and to develop a stratum of rich peasants. The legislative measures to abolish statutory landlordism permitted the payment of huge compensation to landlords and allowed them to retain vast tracts of land. Different kinds of provisions for the right of resumption in tenancy laws led to the eviction of millions of tenants. Loopholes in land ceiling laws helped landlords keep large holdings intact. Millions of acres of surplus land were neither taken over nor distributed to agricultural workers and poor peasants.

Instead of implementing land reform measures and providing facilities for increasing productivity and production to the mass of the peasantry, the Congress governments relied on landlords and rich peasants for agricultural growth, and on their ability to make investments to increase productivity and production.

By contrast, much progress was achieved by the land reform measures introduced in West Bengal, Kerala and Tripura. The strength of the peasant and agricultural worker movements, their struggles and the existence of the Left-led State Governments made the implementation of land reform measures possible in these States. Land reform measures were responsible for noteworthy achievements in advancing the interests of the poor, who constitute the overwhelming majority of the rural people in these states.

With regard to achievements in the West Bengal countryside, Surjya Kanta Mishra and Vikas Rawal (2002) write:

The impact of the wide ranging changes in power structure brought out through land reform and reorganisation of institutions of local government is seen in various indicators of economic and social development in rural West Bengal. Land reform and reorganisation of panchayats created conditions by which West Bengal emerged from a long period of agrarian impasse. In the 1980s, the rate of growth of foodgrain production in the state was the highest among the major foodgrain producing states of India.

Among all States, West Bengal is the highest producer of rice and the second highest producer of potato. Cropping intensity in West Bengal has increased steadily from 136 in 1980-81 to 180 in 2001 — the highest in the country next only to Punjab.

The post-land reform period has been associated with large-scale expansion of irrigation...

West Bengal has also improved its position in allied sectors like fisheries and poultry. For more than a decade, West Bengal has had the highest production of fish and fish seedlings among all the Indian states.

In the 1990s, West Bengal had the highest growth of per capita net state domestic product (NSDP) among all states in the country...

Economic growth in West Bengal has been accompanied by improvement in levels of nutrition and decline in poverty.

The NSS data show that the per capita calorie intake in rural West Bengal increased from 2,027 Kcal in 1987-88 to 2,211 Kcal in 1993-94; in rural India as a whole, by contrast, calorie intake declined from 2,221 Kcal to 2,183 Kcal in the same period. West Bengal is the only state among the major states of India where monthly per capita cereal consumption has increased between 1972-73 and 1993-94. In terms of the head-count ratio, the decline in the proportion of the rural population below the poverty line between 1977-78 and 1997 was the highest in West Bengal among the fifteen most populous states of India; the rural poverty ratio in West Bengal declined by 36.3 percentage points in this period. West Bengal had the second lowest proportion of population below the poverty line among the fourteen most populous states; in 1974, its rank among these fourteen states was eleven.

Indicators of human development also showed steady improvement in the State of West Bengal. The infant mortality ratio (IMR) declined from 92 per thousand in 1977 to 53 per thousand in 1999. Presently, West Bengal has the third lowest IMR and birth rate and the second lowest crude death rate among all Indian states.

Skewed Nature of Distribution of Assets

Assets such as land, agricultural implements, vehicles and transport equipment and pump-sets, tube wells, and other machinery are concentrated in the hands of landlords and rich peasants. So also are financial assets, including access to institutional credit. This skewed distribution of assets has enabled the rich to garner the benefits of productivity and production increases in the rural economy. The concentration of assets in the hands of the rich sections makes the situation more adverse to the interest of the poor.

Facilities Provided by Government

The major portion of government's grants, including subsidies to agricultural inputs, credit facilities, high-yielding seeds, agricultural implements, irrigation and power facilities, and infrastructural facilities (including facilities for storage and cold storage) are cornered by landlords and rich peasants who are the politically influential sections in the bourgeois-landlord political parties in rural India. In areas where the Left and progressive movements are weak, these sections control cooperatives and panchayat raj institutions. The richer sections make use of these institutions for their narrow benefit and against the interest of the overwhelming majority of the peasantry.

Liberalisation policies, based on the retreat of the state and surrender to the capitalist market, have aggravated the crisis in agriculture further. The liberalisation process is biased in favour of the rich and against the interests of the poor, who are mainly agricultural labourers, poor peasants and middle peasants.

Capitalist Development in Agriculture

The economic and agrarian policies pursued by successive governments since Independence have developed capitalist production relations in agriculture and have benefited the rural rich, comprising landlords, capitalist farmers, rich peasants and their allies. The development of capitalist production relations is clearly a major all-India trend but there are regional and sub-regional variations and unevenness. There are regions where capitalism in agriculture has advanced and when commercial agriculture and cash transactions dominate the rural economy; there are regions where old forms of landlordism and tenancy and archaic forms of labour service, servitude and bondage still play an important part in agrarian relations. Caste divisions, caste oppression, gender oppression and the exploitation of the poor by usurers and merchant capital continue unabated all over the country.

Capitalist development in Indian agriculture is not based on a resolute destruction of older forms, but has been super imposed on a swamp of pre-capitalist production relations and forms of social organisation. The development of the "modern" does not preclude the continued existence of the archaic. [Programme of the Communist Party of India (Marxist)]

The development of capitalist relations in the countryside can be seen, *inter alia*, in the proletarianisation of large sections of the rural working masses and the huge increase in the number of agricultural workers as a proportion of the rural population, the differentiation taking place among the peasantry, the large-scale evictions of tenants, production for the market, mechanisation and increased levels of reinvestment of capital in agriculture and agricultural related activity.

TWO PHASES OF CAPITALIST DEVELOPMENT

The post-independence period of capitalist development in agriculture can be broadly divided into two phases — the State-sponsored capitalist development phase and the liberalisation-globalisation phase.

State-Sponsored Capitalist Development Phase

During the State-sponsored capitalist development phase, governments invested in the expansion of irrigation, power, science and technology, transport, communication, storage facilities and so on, and considerable progress was made in these areas. Public investment in science and technology helped develop high-yielding seeds and the production of agricultural implements for increasing productivity and production. Governments also provided subsidies for the purchase of agricultural inputs. Measures were also taken to protect the interests of producers by establishing a minimum support price mechanism in the case of certain crops. Many types of credit facilities were also provided. Priority was given to increasing food grain production and attaining food grain self-sufficiency. Many restrictions were put on the import and export of agricultural commodities in order to protect the domestic market. Even though the lion's share of the benefits of these measures was appropriated by landlords and sections of the rich peasants, these helped the agrarian sector achieve certain levels of growth and development.

The landlords and the richer sections that benefited and gained strength and assets during the phase of State-sponsored capitalist development expected to achieve even more benefits under the globalisation-liberalisation policy regime; hence, most landlord-led peasant organisations supported the globalisation-liberalisation policies.

Liberalisation Phase

The second phase of capitalist development during liberalisation period witnessed a reactionary change in the policies of the state. Many steps have been taken by the state to withdraw from economic affairs and to hand over everything to capitalist market forces to decide. The poor have suffered most during the period of liberalisation. In the place of State-sponsored capitalist development, a strategy of development based on trade came into existence.

Most non-Left state governments are taking steps to reverse the land reform process by relaxing land ceiling and tenancy laws in order to develop the land market. The Government of Karnataka has amended the Land Reform Act of 1961 reintroducing land leasing, allowing non-agriculturists and industrialists to own land and removing land ceiling for aquaculture, horticulture, floriculture and the housing industry. The government of Maharashtra has relaxed restrictions on the conversion of agricultural land to non-agricultural land. The Agriculture Land Ceiling Act has been amended to permit large land holdings. Many non-Left governments have given thousands of hectares of land either on sale or on lease at throwaway prices to multinational companies and domestic monopolies. Wasteland has been given to multinational companies and big business for development and cultivation. Instead of vesting forestland to the adivasis who cultivate it, the Central Government has issued a circular calling for the forcible eviction of lakhs of adivasis from the land. Many non-Left state governments have begun the process of eviction; the Left Front State Governments of Tripura and West Bengal have, however, refused to implement such a measure.

As a result of policies of liberalisation, there have been drastic cuts in public investment in areas such as irrigation, electricity, science and technology, and infrastructural facilities. Public sector investment played a crucial role in the development of irrigation, power, agricultural research, roads, markets and communications. As a result of current policies of liberalisation, investment in agriculture declined from 1.6 per cent of the GDP in 1993-94 to 1.3 per cent of the GDP in 1998-99. This decline was mainly due to the fall in public investment. The share of agriculture and allied sectors in total gross capital formation (GCF) also declined during the period of liberalisation. Contrary to what was claimed at the time of initiating this process, liberalisation policies have been responsible for the decline in investment.

Investments in public health and public education have declined. Privatisation in these spheres has the effect of denying the poor access to facilities for health and education.

The credit-deposit ratio is an important indicator of the degree of involvement of banks in lending. The credit-deposit ratio of rural branches of commercial banks in the country as a whole has declined from 69 per cent in 1991 to 40 per cent in 2001. The decline in the rural credit-deposit ratio has a direct bearing on capital formation in rural India.

The government has reduced subsidies provided to all agricultural inputs, including seeds, water and electricity charges, fertiliser, pesticides, and agricultural implements.

Another consequence of the liberalisation programme is the removal of all restrictions on import and export, other than tariffs. Liberalisation is also leading to the privatisation of water and the creation of water monopolies. The World Bank policy paper on liberalisation of agriculture recommends the creation of "markets in tradable water rights." Through this, water can be diverted from common peasants to corporate super farms. The new agricultural policy of the Karnataka government stands for large corporations and agri-businesses gaining monopoly control over water. The sale of Ganga water to the French company Ondeo Degremont for the Sonia Vihar Water Treatment Plant is another example.

Bio-diversity, which has been the common property of farmers and local communities, is being transformed into the private property of a handful of corporations such as Monsanto, Syngenta, Dupont, Dow, and Bayer. Control over seed means control over agriculture; Monsanto, for instance, clearly uses technology to force farmers into seed dependency. Corporate seeds are creating difficulties for peasants because of the high costs of seed, high dependence on costly inputs (pesticides and herbicides) and high levels of unreliability and crop failure. Cotton cultivators in Maharashtra, Andhra Pradesh and Karnataka complained that the Bt cotton plants were not in fact protected from pests, including the American Bollworm.

The public distribution system is being dismantled. The targeted public distribution system has proved to be an utter failure, and the Antyodaya scheme woefully inadequate in its coverage. Over the last three years, the BJP-led Central government has increased the price of wheat for "above poverty line" households by 85 per cent and the price of rice for "above poverty line" households by 61 per cent. In 2001, the Central government increased the price of wheat for "below poverty line" households by 66 per cent and the price of rice for "below poverty line" households by 62 per cent.

Recurring floods and drought have made serious changes in the environment and the intensity of the adverse effects is increasing year by year. The government is not taking any appropriate remedial measures. Erosion is taking place at an alarming level and a vast extent of land has been made unsuitable for cultivation. The situation is fast deteriorating in most parts of the North-East. Government intervention to combat natural disaster, particularly in the form of compensation and rehabilitation measures, has been very inadequate.

MAJOR EFFECTS OF LIBERALISATION

Decline in Growth Rate of Production and Productivity

Liberalisation has led to an overall decline in the rate of productivity and production in agriculture, particularly with regard to food. The overall growth rate of crop production declined from 3.72 per cent per annum between 1979-80 and 1989-90 to 2.29 per cent per annum between 1989-90 and 1999-2000; and productivity declined from 2.99 per cent per annum to 1.21 per cent per annum over the same periods.

The growth rate of food grain production declined from 3.54 per cent per annum during 1980s to 1.92 per cent per annum in the 1990s. The growth rate of productivity in food grain, which was 3.3 per cent per annum in the 1980s, declined to 1.32 per cent in the 1990s. The rate of growth of production of non-foodgrain crops fell from 4.02 per cent per annum in the 1980s to 2.83 per cent per annum in the 1990s.

Over the period of the ninth five-year plan, the average annual growth rate of food grain production declined to 1.1 per cent, a rate that is lower than the population growth rate between 1991 and 2001, which was 1.9 per cent. The per capita net availability of cereals and pulses, which was 510.1 grams per day in 1991, declined to 414.1 grams per day in 2001.

The performance of the agricultural sector during the ninth five year plan period was below the targeted growth rate. The average annual growth rate achieved during the ninth plan was estimated to be only 2.06 per cent, against a target of 3.9 per cent.

The share of agriculture to the GDP has declined from 61 per cent in 1950-51 to 24.2 per cent in 2001-02. But the population dependent on agriculture has declined marginally from 77 per cent to 69 per cent during this period.

The persistent decline in the relative productivity and income of agricultural worker is a serious disturbing trend. Productivity per worker in non-agricultural activity was 1.8 times that of an agricultural worker in 1950; it had risen to 4.3 times that of an agricultural worker in 1997. Workforce diversification is also not taking place in India.

One of the other disturbing features of current development is the very low employment generating potential of growth in the tertiary sector, which was the fastest growing sector during this period. In the earlier period, agriculture was a fast growing sector but it had a very high employment potential. The result is the increasing concentration of workforce in agriculture and continuous decline of relative productivity, income and living conditions of workers engaged in agriculture.

Price Crash

The liberalisation policies and removal of quantitative restrictions on import of agricultural commodities has led to sharp falls in the prices of many crops in India. According to Dr. Vandana Shiva, "Indian peasants are losing more than Rs. 1,16,200 crores per year due to falling farm prices of staple foods and cash crops." This period also witnessed very wide fluctuations in the prices of many crops. Resource-poor sections of the rural population are worst affected by fluctuations in prices.

Major Sufferers — Agricultural Workers and Poor Peasants

Agricultural workers and poor peasants are the worst sufferers of the policies of the government. A large proportion of them come from dalit, adivasis and other backward classes.

The proportion of landless agricultural workers in the rural population increased from 25 per cent in 1981 to over 40 per cent in 2002. This indicates the growing pauperisation of the poorer sections of the peasantry and their growing alienation from land, which forces them to join the ranks of landless agricultural labourers. The average number of days of work available to an agricultural worker in a year slumped from 123 a year in the 1980s to 78 in the 1990s.

NSS data show that the annual growth rate of rural employment has plunged from 2.03 per cent between 1987-88 and 1993-94 to just 0.58 per cent between 1993-94 and 1999-2000. A large number of retrenched workers from the cities have returned to the villages and have added to the number of rural unemployed. The reduction in rural employment schemes by the Central and State Governments have also adversely affected the employment situation in the countryside. Agricultural workers are subjected to the severest forms of exploitation. The decline in public investment and the reduction of subsidies to agricultural inputs have increased the difficulties of the poor and middle peasants. They also face difficulties in getting institutional credit at affordable rates of interest and hence have to rely on moneylenders who advance loans at exorbitant rates of interest.

Expectations of Market Access Not Fulfilled

The Government of India and many landlord-led peasant organisations were very keen to bring agriculture within the purview of GATT. India, from the very beginning of the Uruguay round of discussions at Punta Del Este in September, 1986, took the stand that "agriculture" should be brought within the purview of GATT. In the present changed world situation, the imperialist countries are exerting pressure on developing countries through international institutions like IMF, World Bank and WTO to accept new conditions and new forms of exploitation. These pressures and the expectations of the bourgeois-landlord sections prompted government of India to take the stand that "agriculture" be brought within the purview of GATT.

The Government of India and the landlord sections expected that the withdrawal of subsidies by the developed countries would result in a significant increase in exports of tradable agricultural goods. Since agriculture is backward and labour-intensive, this would not only provide more employment to agricultural workers but would also help raise their productivity and incomes. Agro-processing industries would also develop and this would lead to a gradual diversification of the labour force, away from agriculture and into non-agricultural occupations. Investment in agriculture and agro-processing industries would also increase, leading to progress and prosperity.

These high expectations were reflected in the inaugural address of Shri Pranab Mukherjee, then Commerce Minister in the Congress Government at the Centre, at a symposium on the impact of the Uruguay Round on agro-exports held in August, 1994:

Coming to the area of trade in agriculture...I would like to emphasise that India was one of the leading developing nations which initiated from the very beginning of the Uruguay round of discussions at Punta Del Este in September 1986, that "Agriculture" should be brought within the purview of GATT. The need for liberalisation in the world trade in agriculture was felt due to extensive subsidisation by the developed countries which led to distortion in the prices of agricultural commodities. As a result, the poor and developing countries like India were finding it difficult to have access to the markets of agricultural products in the developed and developing countries.

A note circulated by the Ministry of Agriculture at a meeting of representatives of peasant organisations and political parties held on September 13, 2000 in New Delhi explained what the Government expected from liberalisation and what happened subsequently.

During the Uruguay round, it was expected that pursuant to the Agreement on Agriculture, distortions in agricultural trade would be reduced and scope for exports of products from developing countries would increase. The anticipated increase in exports of agricultural products from developing countries has not been realised. It was also expected that the contemplated fair trading regime would help the efficient producers in realising higher prices for their products. On the contrary, prices of most agricultural commodities are declining in the world markets. It was anticipated that due to the reduction in domestic support in developed countries, cereal production would shift from developed to developing countries. Empirical evidence, however, shows that there has not been much change in the pattern of world cereals production and exports.

A number of developed countries have continued to provide high domestic support to their agricultural sectors. At best the policies in many developed countries have only been cosmetically altered by shifting the support from one box to another. The level of support can be gauged from the fact that the level of subsidy other than that allowed under the green box, in 1995, in some developed countries was around 30% of the GDP contributed by their agriculture sectors. In absolute terms, the total farm support in the developed countries (OECD) increased by 8% to 363 billion US dollars in 1998. Reportedly in response to a collapse in commodity prices, for the past two years, US Congress has approved a total of $15 billion worth of emergency farm aid during the last three years. The continuation of the high domestic support to agriculture in many developed countries is a cause of concern as they encourage over-production in these countries leading to low levels of international prices.

It is obvious, therefore, that benefits to developing countries in terms of increasing their exports will only occur from complete elimination of export subsidies and substantial reduction in domestic support in the developed countries. In this context, it may be in the interest of India to demand a substantial reduction in the domestic support and export subsidies in developed countries.

Market access to the developed countries is hampered by persistent protection by high tariffs and other barriers. In a recent study of 14 countries, FAO concluded that there was little change in the volume exported or in diversification of products and destination. Tariff peaks continue to block exports from developing countries to the developed world. Tariffs still remain very high in certain sectors, specially, in cereals, sugar and dairy products. Tariff escalation (increase in tariff with successive stages of processing) continues to block exports of value added products from developing countries to the developed countries. Sanitary and phytosanitary (SPS) issues continue to be a major barrier in diversifying exports in horticulture and meat items. Fresh commitments should, therefore, be negotiated to substantially improve market access for products of particular interest to developing countries.

The Government of India claimed that it took a firm position in the ministerial conference of the World Trade Organisation in Cancun in Mexico against the developed countries along with other developing countries and that that led to the stalemate. Nobody disputes the fact that the Government of India took a position demanding reduction of domestic and export subsidies given by the developed countries. But the Government of India and the landlord sections are trying to create an impression that if the developed countries reduce their domestic and export subsidies and peak tariff rates, Indian agricultural commodities will get market access and the situation will change radically in favour of Indian agriculture. The need to demand reduction of subsidies and peak tariff rates of the developed countries cannot be disputed. India can also ask for a "development box" or "food security box" similar to "green box" or "blue box", where the domestic support given is excluded from the calculation of total aggregate measure of support (total AMS).

India — Certain Important Specificities

The strategy of market-oriented development cannot bring development and progress. For the vast majority of the peasantry in India, agriculture continues to be subsistence cultivation and not a commercial venture.

The proportion of poor and middle peasants who hold operational holdings that are not more than 5 acres is very high in India and their numbers are growing with the development of capitalist production relations.

In India, the other important issue to be addressed is the wide regional variation in productivity and production. Productivity levels are higher where irrigation is available. Peasants, particularly poor peasants, living and cultivating in those areas where the productivity level is below the average level suffer many disadvantages. When the minimum support price (MSP) is announced by the Government, there have always been complaints about the low level of the MSP from peasants who live in those areas where the productivity level is comparatively low. Kerala coconut cultivators always complain about the copra MSP announced by the Government of India. The difference in the productivity level of sugarcane, cotton and other cash crops in different regions is another perennial problem.

Another feature of Indian agriculture is that, owing to regional agro-climatic differences, most crops are region-specific. Rubber, coconut and many spices are concentrated in Kerala. Oil seeds are concentrated in dry and rain-fed areas. Rice is concentrated in areas of heavy rainfall and irrigated areas. If edible oil is imported freely because of its relatively low price in the international market, there is no choice for the dry land farmers but to change to other crops. The economies of oilseed mono-crop areas would suffer. Cropping patterns in India are determined by specific agro-climatic situations and there is little room for flexibility in the choice of cropping pattern. In the present circumstances, the peasantry will suffer great difficulties if free trade is allowed.

Productivity and Production Increase — The Crucial Issue

The problems of the peasantry and agriculture can only be solved by increasing productivity and production and ensuring equitable distribution.

The increase in productivity and production is dependent on releasing the creative energy and labour enthusiasm of millions of peasants by breaking up land concentration and traditional systems of exploitation and by distributing land to agricultural labourers and poor peasants. Steps should be taken to protect the interests of the poor from disadvantages hthat flow from the skewed distribution of assets. The major share of concessions and benefits to the countryside should reach the disadvantaged sections.

The productivity level of most crops in India is below the world average level. In the context of an emerging world market, the productivity level of crops in India needs to be raised to world average levels. The interests of the peasantry and agriculture cannot be protected without increasing productivity and production. Serious attempts should be made in this direction. There has been general improvement in the yield of cereals over the years globally. But the improvement in yield achieved by India is very minor. For example, France had the best cereal yield at 7126 kilogram per hectare in 1996-98 followed by Egypt at 6595 kilogram per hectare. Yields were high — more than six tonnes per hectare — in Germany, Japan and the US. The yield of cereals in China was more than twice than that of India. Yields inBrazil and Germany also increased substantially.

Peasantry — Different Sections

Agricultural workers and poor peasants constitute the overwhelming majority of the rural workers in India. They are subjected to ruthless exploitation by landlords, capitalists and by globalisation policies. The middle peasants are exploited by landlords and usurers and the capitalist market controlled by big companies and multinational corporations. Rich peasants constitute an influential section among the peasantry and certain sections among them have gained very much from post-Independence agricultural policies. They employ agricultural workers and have contradictions with them. But they are also adversely affected by constant fluctuations in prices and exploited by the market controlled by big business and MNCs. Even though many of them are inclined to join the capitalist landlord class, the difficulties they face also force them to oppose the policies of the Government. Even landlord sections are facing certain difficulties as a result of steep price falls and price fluctuations under the liberalisation-globalisation regime. They stand for reducing domestic and export subsidies and tariff peaks in the developed countries and other such demands.

Emerging Contradictions

The present situation is characterised by the emergence of two important contradictions. First, the sharp division between the rural rich comprising the landlords, capitalist farmers and their allies on the one hand and the mass of the peasantry mainly agricultural workers, poor peasants and artisans on the other. The second is the growing opposition to imperialist-driven globalisation and liberalisation policies not only from among the mass of the peasantry but also from sections of the rural rich.

A powerful agrarian movement has to be built on alternative policies centred on the interests of agricultural workers and poor peasants who constitute the overwhelming majority of the peasantry and who suffer most due to the growth of capitalism and liberalisation-globalisation policies. The interest of the middle peasants and genuine interest of the rich peasants and the agriculture can only be protected by such alternative policies. The market-driven liberalisation-globalisation process in agriculture is strongly biased towards richer sections, landlords and also prosperous regions.

The richer sections oppose the high subsidies given by developed countries, peak tariff rates and inequitous provisions of the agreement on agriculture as they are against their interests. Some landlord-led organisations that supported liberalisation policies have now begun to oppose them. The agrarian movement should support whoever comes forward to oppose any of the policy measures of liberalisation and globalisation. But some changes in such policies alone cannot solve the real problems faced by Indian agriculture and the overwhelming majority of the peasantry in the country. An alternative that is centred on the interests of agricultural workers and poor peasants and also works in defence of the genuine interests of middle and rich peasants alone can find a solution to all problems. It is also necessary to wean away these sections of the peasantry from the ideological influence of landlords and other richer sections to build a powerful agrarian movement based on the genuine interests of the peasantry, agriculture and the country.

A powerful all-India agrarian movement can be built only through struggles against the policies of the state, which is the organ of the class rule of the bourgeoisie and landlords led by the big bourgeoisie who are increasingly collaborating with foreign finance capital in pursuit of the capitalist path of development.

Alternative Policies

The present trade-oriented strategy of agricultural development should be rejected and an alternative policy of development that stands for active intervention by the state to increase productivity and production in agriculture should be implemented in its place. The great unevenness in agricultural development — both spatial and sectional — in India creates great difficulties for the poor and for peasants living in backward areas. The state should make appropriate interventions in developing backward areas. The persistence of poverty and unemployment in rural India can only be solved by increasing productivity and production in agriculture and rapid industrialisation. If food security is to be ensured over time, production and yields must increase. Given the emerging globalisation of agricultural markets, increases in yield are the starting point for the protection of the peasantry and agriculture in India. The technological changes that are required to upgrade agricultural productivity require sustained and multi-dimensional state intervention.

An alternative agrarian and agricultural policy should include the following measures:

1. Land reforms must be implemented and land distributed to agricultural workers and poor peasants in all states where such measures have not been implemented. The steps to reverse land reform measures must be resisted.
2. The state must ensure minimum wages and social security measures such as provident fund, pension, and workers' compensation, in order to improve the standards of living of agricultural workers. The economic and social conditions of agricultural workers should get adequate attention and priority. The central government should pass a comprehensive law for agricultural workers in order to ensure minimum wages and provide protective social security.
3. Government should make investments to expand and upgrade irrigation, electricity, agricultural science and technology (including technological applications for small holdings), transport and communication facilities in the countryside, agricultural processing centres, and storage units (including cold storage facilities). With regard to agriculture and rural development, the policy of retreat by the state must decisively be reversed.
4. Steps should be taken to ensure food security for the people.

5. Government should make available adequate credit at low rates of interest. This is essential if peasant investment in agriculture is to be sustained, and is essential to provide some protection to the rural poor from moneylender exploitation.
6. Government should provide subsidies to agricultural inputs to protect the interest of the poorer sections among the peasantry. Additional subsidies should be given to peasants farming in backward areas. Appropriate types of peasants committees should be constituted to ensure that the benefits of subsidies reach genuine beneficiaries.

Share of Agriculture and Allied Sector in total GCF (%)

Year	Public Sector	Private Sector	Total
1987-88	10.1	13.2	11.7
1988-89	8.8	9.7	9.3
1989-90	7.5	9.1	8.4
1990-91	7.1	11.9	9.9
1991-92	6.6	9.9	8.7
1992-93	6.7	10.5	9.1
1993-94	6.9	9.4	8.4
1994-95	6.7	7.7	7.3
1995-96	7.1	5.9	6.2
1996-97	7.0	7.5	7.4
1997-98	6.2	7.5	7.1
1998-99	5.7	7.8	7.2
1999-2000	5.1	8.2	7.2
2000-01*	4.9	8.2	7.1

* Quick estimates

Source: Central Statistical Organisation, New Delhi.

7. A comprehensive crop insurance scheme covering all crops should be implemented.
8. Government of India should take a firm stand on WTO related issues and should try to rally other developing countries to make making appropriate changes to eliminate unjust and iniquitous provisions in the agreement on agriculture in WTO. India should ask for:

a) Reduction of domestic and export subsidies of developed countries.
b) Elimination of all boxes like green, blue and amber. Fixed allowable minimum subsidy either on the basis of per acre or per capita.
c) Reintroduction of quantitative restrictions in the case of those commodities that get domestic or export subsidies above the allowable minimum.
d) Reduction of tariff peaks.

Changes in the distribution of land holdings in India from 1980-81 to 1990-91 (Numbers in 000' area in 000' hectare)

Major size classes	Number			Area		
	1980-81	1985-86	1990-91	1980-81	1985-86	1990-91
Marginal (below 1 hectare)	50,122 (56.4)	56,147 (57.8)	63,389 (59.4)	19,735 (12.0)	22,042 (13.4)	24,894 (15.1)
Small (1 to 2 hectare)	16,072 (18.1)	17,922 (18.4)	20,092 (18.8)	23,169 (14.2)	25,708 (15.6)	28,827 (17.4)
Semi-medium (2 to 4 hectare)	12,455 (14.0)	13,252 (13.6)	13,923 (13.1)	34,645 (21.2)	36,666 (22.3)	38,375 (23.2)
Medium (4 to 10 hectare)	8,068 (9.1)	7,916 (8.2)	7,580 (7.1)	48,470 (29.6)	47,144 (28.6)	44,752 (27.0)
Large (10 hectare and above)	2,166 (2.4)	1,918 (2.0)	1,654 (1.6)	37,705 (23.0)	33,002 (20.1)	28,659 (17.3)
All size classes	88,883 (100)	97,155 (100)	1,06,637 (100)	1,63,724 (100)	1,64,562 (100)	1,65,507 (100)

9. The government should intervene to protect the peasantry from sudden falls in the prices of agricultural commodities.
10. The public distribution system should be universalised and strengthened.
11. Panchayat institutions should be strengthened, voluntary cooperatives should be organised to help the peasantry in production, marketing, and primary processing.
12. Agro-based industries should be developed. The entry of MNCs into agriculture should be regulated.
13. The state must take adequate measures to protect the environment, regenerate degraded areas, prevent soil erosion, prevent and control pollution, and conserve forests, wild life, flora and fauna. The state must promote environmental research, the dissemination of environmental information and the creation of environmental awareness among the people.

Comparative yield of principal crops (Kg per hectare)

Country	Paddy	Wheat	Maize	Groundnut	Sugarcane
India	2929	2583	1667	913	68012
China	6321	3969	4880	2799	85294
Japan	6414			2336	
USA	6622	2872	8398	3038	80787
Indonesia	4261		2646	1523	
Canada		2591	7974		
Vietnam	4105			1435	
World average	3845	2711	4313	1336	65689
Rank of India in production in the world	Second after China	Second after China	Accounts for only little over 4% of world's production	Second after China	Second after China

Source: Agriculture at a glance, 2002, Ministry of Agriculture.

Improvement in Yield (kg/hectare)

Country	1979-81	1996-98
France	4700	7126
Egypt	4053	6595
S. Korea	4986	6450
Germany	4166	6366
Japan	5252	6017
Australia	4143	5693
USA	4151	5380
China	3027	4821
Brazil	2047	4081
Indonesia	2837	3916
Vietnam	2049	3754
Argentina	2183	3284
Sri Lanka	2462	3103
Malaysia	2828	3065
Thailand	1911	2466
India	1324	2200
Australia	1321	1973
Cuba	2458	1973

Source: World Development Indicators, 2000

*Paper Presented At The All-India Conference On Agriculture And Rural Society In Contemporary India, Barddhaman, December 17 To 20, 2003

7

Transition and Economic Development in India

While much of India is exceedingly poor—with a per capita GDP of close to $3,000 in purchasing power parity—Indians play an increasing part in global commerce. Expatriate Indians have a significant presence in Silicon Valley, and India itself now has a thriving software industry based in Bangalore in southern India. The phenomenal growth of the information technology industry in India is documented in the article by N.R. Narayana Murthy, the founder of Infosys, one of the big software firms that pioneered the speedy development of that industry. '[I]ndia's annual software exports increased from US$24 million in 1985 to US$164 million in 1992 and to US$7.8 billion in March 2002. Similarly, India's domestic software sales increased from US$140 million in 1992 to US$2.45 billion in 2001-2002' (p. 217). Murthy attributes this growth partly to reforms initiated after the balance of payment crisis of 1991.

An article by Nirvikar Singh is a more detailed analysis of the contribution of the information technology sector to the growth of the Indian economy. At present, the information technology sector is only 3 per cent of the economy. For Singh, this means that there is further room fcr accelerated growth. The development of the software industry leads one to wonder about the potential for growth of the hardware sector. In this, however, the prospects are not as rosy. As Singh points out, the development of hardware manufacturing requires infrastructure sorely missing in India. Will the development of the information technology industry be sufficient to generate the growth and development that are needed for the Indian economy as a whole? The author discusses the potential for spillover effects and the development of IT-enabled industries. He argues that policies targeted to benefit the software industry may be misplaced. A better way to help the economy would be policies that include sensible labour laws, generate competition through proper regulation and provide basic infrastructure.

The final segment of this collection includes an essay on the informal economy by Barbara Harriss-White. A large segment of Indian society functions outside the formal economy. According to the 1991 census, 88 per cent of Indian society is part of this informal sector, mostly located in the rural areas and largely driven by an intermediate class of small business owners and enterprises. '[I]n relation to big business [the intermediate classes] appear small and dispersed, but in relation to labor they present a mighty front; they are the "masters of the countryside"' (p. 278). The power of the intermediate classes means that subsidies for fertilizers, food, agricultural electricity and credit have proved extremely hard to reduce. There is also widespread evasion of taxes by the intermediate classes and widespread flouting of the constitutional entitlements of labour. The author's conclusion is that in order for development to reach higher levels, the legitimacy of the state will need to be renegotiated.

Finally, R. Jhabvala and R. Kanbur study the impact of globalisation and economic reform on unskilled workers by focusing on the actions of Self Employed Women's Association in India (SEWA), a non-governmental organisation. It is an extremely interesting case study of how a group of unskilled workers, all women, have organised themselves to cope with the changes brought about by increasing liberalisation and globalisation. Unskilled workers face a relative decline in wages, increasing risk and vulnerability, and a decline in bargaining power. Increasing the skill levels of the poor and offering insurance to manage the risks they face are remedies. SEWA has been able to do some of this on its own, but the authors suggest that the government should intervene more directly to help such organisations, as well as to improve their legal and regulatory environment.

This collection should be of immense value not only to those who are curious about the Indian economy but also to scholars seriously interested in development. The Indian experience has been a rich one with many lessons to be learned. Even with its very wide range of subjects, it is impossible to cover every aspect of the development process in one volume, especially for an economy as vast and complex as India's. One important topic left out is the role of microfinance in serving the credit needs of the poor. The service sector has also been a major contributor to the rapid growth of the Indian economy in the last decade. Tourism has been growing as rapidly, if not more rapidly, than the information technology sector. For the first time during the holiday season of 2005, Goa on India's west coast has overtaken the USA as a destination for vacationers from the UK. So, indeed, the growth in the Indian economy has been much more broad-based than would be evident from the collection of essays in this excellent volume.

CLASS ANALYSIS OF INDIAN AGRICULTURE

No doubt, a growth in "consumption" and "distribution" (i.e. growth in the home market) is an indication of the growth of capitalist relations in production but the important question is that what is its impact on production relations, and, even with this growth of commodity production which of the two modes of production (i.e., capitalist and feudal) is predominant in Indian agriculture ? This can only be done by taking an overall picture of production relations and not merely by picturing the growth in "consumption" and "distribution". Lenin has said, *"that the economists who have discoursed at length on the inadequate atten-tion paid by the classical economists to 'distribution' and 'consumption' have not been able to give the slightest explana-tion of the most fundamental problems of 'distribution' and 'consumption'. That is understandable, for one cannot even dis-cuss 'consumption' unless one understands the process of the reproduction of the total social capital and of the replacement of the various parts of the social product... It is not with 'production' that political economy deals, but with the social relations of men in production, with the social system of produc-tion. Once these social relations have been analysed, the place in production of every class, and consequently, the share they get of the national consumption is thereby defined."* (Development of Capitalism in Russia........Lenin)

Capitalism does not merely entail commodity production. As Lenin has said, *"Capitalism is commodity production at its highest stage of development, when labour power itself becomes a commodi-ty."* (Imperialism the Highest Stage of Capitalism.... Lenin) Besides, when the CRC, Communist League, etc., speak of increasing "consumption" and "distribution" they forget Marx's basic theory of realisation which states that :

"According to the general law of capitalist production, constant capital grows faster than variable capital" and that *"capitalist production, and, consequently, the home market, grow not so much on account of articles of consumption as on account of the means of produc-tion. In other words, the increase in means of production out-strips the increase in articles of consumption."* (Development of Capitalism in Russia.......Lenin)

In other words, while dealing with the growth of commodity production and capi-talist relations in India, there is a need to picture the nature of the growth - in which spheres is it confined to, and to what extent is it confined to consumer items and to what extent capi-tal items. Is it the sphere of productive consumption that is increasing in agriculture or is it the sphere of consumer expend-iture that has increased ?

While analysing the mode of production in Indian agriculture it is necessary to do so based on the laws of capitalist production as discovered by Marx and further developed by Lenin. These scientific laws of capitalism and imperialism are not time bound.

They neither get out-dated nor change from country to country. The conditions may change in time and space but the basic laws of capitalism and imperialism remain unchanged. To understand the mode of production in India it is no use negating these laws, as the CRC does, but utilise them to understand the mode of produc-tion in India.

Regarding the question of capitalism in agriculture an important principle outlined by Lenin was that, *"it is the development of capitalism in the manufacturing industry that is the main force which gives rise to, and develops, capitalism in agriculture."*

Also, while talking of the transformation of pre-capitalist forms of production to capitalism, Lenin has said that :

"the socialisation of labour by capitalism is manifested in the following process :

Firstly, the very growth of commodity production destroys the scattered condition of small economic units that is characteris-tic of natural economy and draws together the small local markets into an enormous national (and then world) market. Production of oneself is transferred into production for the whole of society: and the greater the development of capitalism, the stronger becomes the contradiction between this collective character of production and individual character of appropriation.

Secondly, capitalism replaces the former scattered production by an unprecedented4d concentration both in agriculture and in industry.

Thirdly, capitalism eliminates the forms of personal dependence that constituted on inalienable component of preceding systems of economy ...

Fourthly, capitalism necessarily creates mobility of the popula-tion, something not required by previous systems of social econo-my and impossible under them on anything like a large scale.

Fifthly, capitalism constantly reduces the proportion of the population engaged in agriculture and increases the number of large industrial centres.

Sixthly, capitalist society increases the population's need for association, for organisation, and lends these organisations a character distinct from those of former times. While breaking down the narrow, local social-estate associations of medieval society and creating fierce competition, capitalism at the same time splits the whole society into large groups of persons occu-pying different positions in production and gives a tremendous impetus to organisation within each such group.

Seventhly, all the above mentioned changes effected in the old economic system by capitalism inevitably lead also to a change in the mentality of the population. The spasmodic character of economic development, the rapid transformation of the methods of production and the enormous concentration of production, the disappearance of all forms of personal dependence and patriar-chalism in relationships, the mobility of population, the influ-ence of the big industrial centres, etc., all this cannot but lead to a profound change in the very character of the producers." (Development of Capitalism in Russia, by VI Lenin) With this brief introduction on the approach to the problem, let us now turn to study the relations of production in Indian agri-culture.

While studying the relations of production in Indian agriculture we will examine it under two heads : a) Growth of Commodity Production, and b) Disintegration of the peasantry.

A. GROWTH OF COMMODITY PRODUCTION

The growth of commodity production in agriculture can be assessed by the extent of utilisation of inputs and the growth of outputs produced for the market. The 'green revolution', which was intro-duced by the World Bank in Third World countries in the mid-1960s, was part of the imperialist policy to penetrate the coun-tryside for markets.

This 'green revolution' has led to certain changes in agriculture which must be analysed. This section will be classified under the following sub heads : 1) 'Green Revolu-tion', 2) Inputs, 3) Credit, 4) Output and Productivity Trends, and 5) Utilisation of Surplus.

1) Green Revolution ....

2) Inputs

Though irrigation may have increased, out of the 350 million acres of net cropped area, barely 50 million acres, or 14%, is cultivated more than once a year (VI Five Year Plan figures). If one crop takes on an average four months, 86% of India's agricul-tural land remains idle for 66% of the time per year. To put it differently, nearly 60% of the cropped area is not at all used. If agriculture was to be put on a capitalist footing and capital-ist farming was to predominate in India, there would not be such a gigantic wastage of land utilisation and such a slow growth of irrigation facilities.

But besides the modernisaation, the number of wooden ploughs increased from 37.5 million in 1956 to 43 million in 1966 and further to 44.5 million (i.e. during the green revolution period) in 1972

3. Credit

It is thus clear that the stranglehold of usurious capital still dominates all sections of the rural populace. But usurious is an indication of pre-capitalist, feudal relations of production, which sucks up the surplus and prevents it from seeking produc-tive channels. Besides, this usurious capital is increasing day by day. Marx has said that, *"Usurer's capital as the characteris-tic form of interest bearing capital corresponds to the predomi-nance of small scale production of the self-employed peasant and small master craftsman", "Usury centralises money wealth where the means of production are dispersed. It does not alter the mode of production, but attaches itself firmly to it like a parasite and makes it wretched. It sucks out its blood, enervates it and compels reproduction to proceed under even more pitiable condi-tions."* (Karl Marx, Vol. III, Chapter 36, pp594)

.............Also with a large section of the peasantry having turned to HYV, (and now unable to turn back to the traditional varieties) and caught in a crisis with diminishing returns for their output, the peasants are dependent on large capital inputs each year which they are now unable to finance. This is forcing them to seek larger and larger loans. With the peasantry unable to even pay their interest on their cooperative loans, they are turning more and more back to the moneylender. This trend is bound to increase enormously in the coming years as the crisis in agriculture deepens.

4. Output and Productivity trends

In fact, the annual rate of growth of production of all agricul-tural crops actually dropped in the second period; while it was 3.2% in the 1951-52 to 1964-65 period, it was just 2.6% in the 1964-65 to 1983-84 period. This was because the increase in area under crops increased phenomenally in the first period at an annual rate of 1.7%, while in the second period the annual in-crease was just 0.4%. In other words, much better results in agricultural production could have been achieved by increasing the area under cropping than by introducing HYV.

The economic and scientific research foundation has estimated that soil erosion and deforestation has been so acute that India looses about 1% of its cultivable land every year to deserts. It is said that out of a total of 306 million hectares of cultivable land, 145 million hectares are either threatened with erosion or badly in need of soil and water conservation measures.

Generally, it can be summed up that production and yield of cropping has increased, with small increase towards cash crops, indicating some growth in capitalist relations. This is indicated through the change in some pockets of agriculture to capitalist farming and growth in commodity relations in large tracts of semi-feudal production, which continue their semi-feudal exist-ence with enhanced contradictions caused by this lopsided growth of commodity relations within it. Productivity trends continue to indicate a backward mode of production with large farms being the least productive. Also with the increase in value of output being nowhere commensurate with the increase in value of inputs, it is clear that the growth in commodity production is lopsided, with this growth merely facilitating the dumping of commodities of the imperialists and comprador big bourgeoisie into the rural sector without significant returns to the farmer. This has led to a new set of contradictions in agriculture which are bound to intensi-fy.

5) Utilisation of Surplus

This, in fact, is one of the factors in the determination of capitalist growth, as a fundamental law of capitalism is that constant capital (i.e., production of means of production) must grow at an increasingly fast rate. Therefore, in capitalist farming the kulak must reinvest his surplus in the farm in im-proved technology. Thereby he would increase the productivity of his farm and the surplus value extracted from it.

Besides this low generation of surplus in agriculture, even that generated goes primarily to other spheres. With the return on money-lending and trading far more profitable, a large proportion of this surplus is not re-invested in agriculture, instead finds its way in such spheres of activity. In fact, as long as usury continues to dominate the countryside as a most profitable sphere of investment, it will restrict the growth of capitalist develop-ment. Of course, with the inception of the 'green revolution' things have somewhat changed; but the limited extent to which it has generated capitalist farming can be seen from the poor levels of capital investment in agriculture taken as a whole.

Also 'capital goods' imports into the rural areas also began to increase, though at a much slower rate, in the same period.

SUMMARY

From the entire section on the growth of commodity production it will be seen that this growth is definite .... whether in the sphere of inputs, outputs, credits utilisation of surplus - all sections have to some extent been affected, thereby having an impact on the production relations in the rural economy. But the question is to what extent has it been able to change the rela-tions of production in agriculture to capitalist ? To get a comprehensive answer to this question it is necessary to analyse the differentiation of the peasantry and all its related aspects. Meanwhile, it must be remembered that even by 1971-1972 only roughly 35% of grain production actually comes into the market. The bulk, over 65%, is consumed by the peasantry without ever reaching the market. This proportion of marketed grains has not changed perceptively in the last decade.

B. DIFFERENTIATITION OF THE PEASANTRY

The differentiation of the peasantry means the break up of the rural populace into a rural bourgeoisie and a rural proletariat. Lenin has said that for Russia, *"two main lines of its develop-ment and outcome are objectively possible : either the old land-lord economy, bound as it is by thousands of threads of serfdom, is retained and turns slowly into purely capitalist, 'junker' economy. The basis of the final transition from labour service to capitalism is the internal metamorphosis of feudalist landlord economy. The entire agrarian system of the state becomes capital-ist and for a long time retains feudalist features. Or the old landlord economy is broken up by revolution, which destroys all the relics of serfdom, and large land ownership in the first place. The basis of the final transition from labour service to capitalism is the free development of small peasant farming, which has received a tremendous impetus as a result of the expro-priation of the landlord estates in the interests of the peasant-ry. The entire agrarian system becomes capitalist, for the more completely the vestiges of serfdom are destroyed the more rapidly does the differentiation of the peasantry proceed."* (Preface to the II edition of Development of Capitalism in Russia..... Lenin)

To do so (ie. examine the differentiation of the peasantry and the growth of commodity production resulting from it) we will analyse it under the following sub-heads :

1) Rural-Urban Divide,
2) Agricultural Labour Vs. Kulak
3) Land Holding-Land Reforms
4) Household Industry
5) Home Market

1. Rural-Urban Divide

Lenin has said that, "the development of commodity economy means the divorcement of an ever-growing part of the population from agriculture, i.e., the growth of the industrial population at the expense of the agricultural population." In explaining the entire process, Marx clarifies, "it is in the nature of capitalist production to continually reduce the agricultural population as compared to the non-agricultural, because in industry the in-crease of constant capital in relation to variable capital goes hand in hand with an absolute increase, though relative de-crease, in variable capital, on the other hand in agriculture the variable capital required for the exploitation of certain plot of land decreases absolutely; it can thus only increase to the extent that new land is taken into cultivation, but this again requires as a pre-requisite a still great growth of the non-agricultural population."

2. Agricultural Labourers vs Kulaks

Also, the productivity of farm workers in India, on the average, is very low and the instruments of production little developed. And very often, the labourer is not divorced form the instruments of production as he has to use his own implement while working on someone else's farm. One important aspect of capitalism is that wage power must be divorced from the instruments of production.

So though a greater number of the peasantry may be disposed from the means of production and land and though wage labour may have increased with the 'green revolution', the bulk of the labour does not take on the form of regular free wage labour. So the clear differentiation of a rural proletariat and a rural bour-geoisie is yet to coalesce, except in few pockets.

The result will be greater amount of labour turning into a commodity coupled withy further impoverishment of the peasantry. Yet this sale of labour power will more take the form of simple commodity produc-tion rather than capitalist production. For, say, a middle peas-ant is unable to carry out his farming operations with family labour and so hires three to four workers during the season.

Yet, if the bulk of the produce his family consumes, his exploi-tation of wage labour gives him no surplus value and so takes the form of simple commodity production even if his produce is a cash crop - say cotton. Much of the labour power sold in the Indian countryside takes this form, and though because of it commodity production expands, the home market grows and there is some change in agrarian relations, but fundamentally it takes the form of simple commodity production with a growth of small-scale farming without generating capitalism or a rural bourgeoisie. Also, it will be observed that labour-service in the form of share cropping and family cooperation co-exist and continue together with the emerging wage-labour system in Indian agricul-ture.

As regards the extent of growth of the rural bourgeoisie (Kulak) only a general picture can be got from the extent of capital formation in agriculture and the extent of the absorption of the means of production in agriculture. These various estimations are given in other sections. The particular extent of its existence can only be ascertained by specific area studies. The following are some general guidelines to assess whether a farmer comes within the category of rural bourgeoisie or continues as a rich peasant-landlord.

a) The bulk of the surplus must be extracted by the exploitation of labour power. He must employ more wage labour than the labour of his family.

b) The major part of the surplus value generated must be utilised in developing the means of production (machinery) and not in usury, trading, purchase of land, etc.

c) Labour power used must be free wage labour and the labourer should be divorced from the instruments of production - i.e., the implements for working should not be brought by the labourer, but instead provided by the owner of the farm. Payment should be in cash and not in kind.

d) The capitalist farmer must not give his land on rent, on the contrary, he may lease in land.

e) The capitalist farmer must produce for sale. So that bulk of his produce must be sold in the market and not consumed by him and his family.

f) The capitalist farmer should produce for profit - i.e., he must accumulate capital continuously. He must plan his production in such a manner that it enhances profit through agricultural operations.

By taking these six criteria into account we can roughly estimate the class character of the farmer - whether he belongs to the category of rural bourgeoisie or rich peasant-landlord class. Studies may show that a large number of such farmers may be found more to be in a transitional stage not having fully coalesced into a rural bourgeoisie and thereby meeting only some of the above mentioned six criteria.

3) Land holdings - land reforms

A full ten years after the so-called 'green revolution' the state of operational holdings in the country remained basically the same.

So this parcellisation and fragmentation of land is continuing apace, and if, within it, a certain amount of commodity produc-tion has been introduced with the help of HYV, cooperative loans, etc., it can lead to only simple commodity production and not capitalist relations of production.

Thus it will be seen that the main trend in agricultural land holding is continuing concentration of land in the hands of a few coupled with extensive small-scale farming. In between lie a class of rich peasantry (15-25 acres) who comprise 3.4% of the households and own 16.2% of land whose number has remained virtu-ally static over the years. It is this section that has, of late, been more vocal as a result of 'green revolution.'

This exposes the glaring existence of labour service in Indian agriculture. The continuing prevalence of tenancy, share-cropping, etc., is a further indication of the perseverance of feudal mode of production. The estimates of cash rent and rent-in-king may not be too accu-rate and estimates vary from study to study. Other estimates put the rent-in-kind around 25%. But, one important point is that the content of the rent, whether in cash or kind, takes the form of pre-capitalist rent and not capitalist rent, as it involves distribution of surplus product and not of surplus value. The very tenancy acts themselves fix the proportion of distribution of the product between the landlord and serf. If it was to be capitalist rent the distribution would be assessed on the profit generated and not on the basis of the whole product.

4) Household Industry

Caste-based hereditary labour, according to the 1971 Census, increased over that of 1961 and stood at : barbers 7.11 lakhs, dhobies 9.47 lakhs, fishermen 6.01 lakhs, weavers and spinners 34.03 lakhs, shoemakers 5.97 lakhs, carpenters 12.64 lakhs, blacksmiths 10 lakhs, jewelry and precious metal workers (incl. goldsmiths) 5.76 lakhs, masons 15.35 lakhs, potters 9.66 lakhs, and tanners 0.53 lakhs.

If capitalist industry were to grow and break feudal relations it should first and foremost oust these ancient and backward forms of production and service. But these continue to co-exist, and even expand together with industrial growth. Also a differentia-tion of the peasantry would simultaneously see the collapse of the household sector, coupled with the growth of modern indige-nous industry. Lenin has said that, *"Domestic industries are a necessary adjunct of natural economy, remnants of which are nearly always retained where there is a small peasantry."* (Development of Capitalism in Russia......Lenin)

In India, not only does the caste based household labour contin-ue, but the household industries are propped up by the state and comprador big bourgeoisie. For example, in the year 1981-82 the total grants plus outstanding loans to the Khadi and Village industries was Rs. 338 crore - Khadi Rs. 208 crore and village industries Rs. 130 crore -given by the government. The very fact that this sector not only continues to exist but is propped up even in the face of expanding industrial production, is a clear example of the compromise struck by imperialism and the comprador big bourgeoisie with the backward pre-capitalist modes of produc-tion which they use as a social base for their existence.

Home market

The differentiation of the peasantry will lead to an expansion of the home market.

In India, as already mentioned, though there has been a continu-ous increase in food grain production, the percentage finding its way into the market has been static at around 30% to 35%....................A more recent estimate put that no more than the top 10% of th land-owning peasant household has able to produce marketable surpluses on any scale.

Also it will be found that the gross domestic capital formation as a percentage of the total has been virtually static in the decade of 1970s. It increased from 18% of the total (compared to registered manufacturing of 20%) in 1970-71 to 19% of the total (compared to registered manufacturing of 24%) in 1980-81. With little growth in percentage of capital formation in the rural areas it would reflect little growth in capitalist relations there during that period.

Also, during the entire period of the 1950s and 1960s there was a massive shrinking of the percentage of per capita consumer ex-penditure spent on industrial goods in the rural areas (and also urban areas). In other words there was a relative shrinking of the home market in that period with a larger and larger amount of the families' incomes being devoted to the necessities of life. There was a perceptible reversal of this process in the rural areas from 1968-69 indicating some growth of market from the time of the 'green revolution.'

SUMMARY

This entire section shows that there has been a very slight differentiation of the peasantry. The basic forms of pre-capitalist production relations has remained intact, with land concentration continuing as before and small scale petty farming dominating the bulk of the rural population. It is within this existing framework that commodity production and the agrarian market has been increased by the imperialists through the 'green revolution', thereby enhancing capitalist relations within the semi-feudal framework.

Thus, from this section we find that the 'green revolution' period has definitely witnessed a rise in commodity production. This is reflected in :

a growth of the urban population, growth of the rural market, growth of banking and institutional finance in the rural areas, a slight increase in constant capital in agriculture,

the widespread prevalence of HYV types, and a slight growth in productivity.

But at the same time all the pre-capitalist institutions continue to co-exist, if not increase. So we find, no change in land holdings and the growth of backward small scale farming, the traditional wooden and iron ploughs and bullock carts contin-ued, widespread prevalence of household industry, the bulk of foodgrain production still being produced for con-sumption and not for market, the bulk of the surplus still not going to increase constant capital but for usury, trading, etc.

no scientific utilisation of agriculture with productivity de-creasing with increasing size of farm, the continuing prevalence of tenancy, share-cropping, etc. the bulk of the population continuing to be dependent on land, little differentiation of the peasantry with little growth in the agricultural labour, With the labour generated continuing to be not free labour but attached and that which is free labour is being more of a season-al type, and the continuing caste-based division of labour of handicrafts and services.

It is clear from this that the growth in capitalist relations has merely been super-imposed on the existing semi-feudal structure and has not proceeded to smash it.

No doubt, this increased capitalist penetration will enhance the contradictions within the rural economy, undermining to some extent traditional feudal relations and have its own impact on the class antagonisms in the countryside. One effect of this has been the mass upsurges of the middle and rich peasantry in many HYV areas. Though also some pockets of bourgeois farming will have developed (its limit can be seen in the excessively low increase in constant capital and particularly of the means of production, within Indian agriculture taken as a whole) the major impact of the 'green revolution' has been to enhance simple commodity production and not capitalist production.

Simple commodity production has existed since slave society, while capitalism, as already mentioned, *"is commodity production at its highest stage of development, when labour power itself becomes a commodity."* Simple commodity production is present when constant capital remains static with little or no surplus value being generated with the bulk of the produce being utilised to sustain the person and his family. Capitalist production, on the other hand, entails a continuing increase in constant capital and a generation of surplus value which is utilised to enhance the means of production.

So, whether it is a petty farmer, utilising even some limited wage labour, or whether it is a handicrafts man or some household industry, or whether it may be any other petty producer - they may all produce commodities but neither do they generate surplus value not do they develop or increase constant capital. This form of simple commodity production can continue for generations with out having much impact on the social relations of production. Capitalist production, on the other hand, revolutionises the mode of production and by a continuous increase of the constant capi-tal leads to the ruination and destruction of the small producer. The primary impact of the 'green revolution', except developing a rural bourgeoisie in certain pockets, has led to the growth of simple commodity production on the lakhs of small fragmented farms through the widespread utilisation of HYV.

Also, the enhancement in sale of labour power utilised in simple commodity production may extend the market but will not be a measure of the growth of capitalist relations. In India, though the percentage of agricultural labourers may not have increased much in the last decade, there has been an increase in sale of wage labour primarily because of the vast expansion of uneconomi-cal small holdings of the poor and even lower middle peasantry. But the bulk of this wage labour is involved in simple commodity production with little effect on the growth of capitalist rela-tions.

Finally, it is manufacturing industry that is the main force which develops capitalism in agriculture. Though industry has been growing, it has been doing so as an off-shoot of finance capital and therefore has been unable to generate that employment that is necessary to absorb any differentiation of the peasantry that may take place in the countryside. So the growth of the manufacturing industry within the imperialist framework, has not acted as that force to develop capitalism in agriculture. On the contrary, we have seen, as with the house hold industry, it even allies itself with the pre-capitalist sectors of production. The result of such warped industrial growth is a massive unemployment in the urban areas together with even larger under-employment in the rural areas.

Thus, the overall agrarian picture in India continues to be semi-feudal within which there has been a certain growth of capital-ist relations.

8

Shifts in the Distribution of the Productive Forces

INTRODUCTION

The structural changes in the industries of any large economic system of any kind embrace an intricate bundle of extremely varied but inseparably linked phenomenon and processes. In summarised form they display many features of the main development trends of society's productive powers in any one historical interval or another. Their generalised characteristic is a high degree of abstraction, necessary for scientific analysis, uniting the most diverse types of man's productive activity in a single whole. As Marx wrote:

Production in general is an abstraction, but a sensible abstraction in so far as it actually emphasises and defines the common aspects and thus avoids repetition. Yet this *general* concept, or the common aspect which has been brought to light by comparison, is itself a multifarious compound comprising divergent categories.

The indicators mentioned in the preceding chapters of the resultant development trends in the postwar capitalist economy also reflect a high degree of generalisation. Nevertheless they are extremely important for a combined study of the real course of events. A scientific methodology for analysing broad social processes, including economic ones, requires us without fail 'to proceed from concrete realities, not from abstract postulates'.

In this connection Marxism-Leninism has always attached paramount importance to developing methods of making qualitative, i.e. statistical, estimates of socio-economic processes.

Lenin, stressing the need for a concrete approach to 104 study of the main forms of capitalism's development in industry, wrote that one must examine the development of any particular form, after bringing out its essence and distinguishing features, by means of 'properly compiled statistics'. He made the same demand in regard to agriculture, for study of which we needed 'a picture of the process *as a whole*, with all the trends taken into account and summed up in the form of a resultant'. Only in that way can the decisive tendencies of economic development be concretely determined and the main point in them separated from the chance and secondary.

In his article 'Statistics and Sociology', written in 1917, Lenin stressed that the most widely used, and most fallacious, method in the realm of social phenomena is to tear out *individual* minor facts and juggle with examples. Selecting chance examples presents no difficulty at all, but is of no value, or of purely negative value, for in each individual case everything hinges on the historically concrete situation. Facts, if we lake them in their *entirety*, in their *interconnection*, are not only stubborn things, but undoubtedly proof-bearing things. Minor facts, if taken out of their entirety, out of their interconnection, if they are arbitrarily selected and torn out of context, are merely things for juggling, or even worse.

Hence a problem of vital interest for the study of social systems of any size, namely that of analysing the interaction of the whole and the part, inevitably arises, because the universal cannot exist otherwise than 'in the individual and through the individual'.

This problem becomes particularly acute when we are trying to elucidate the dynamics of the long-term tendencies in the distribution of the productive forces within such a broad system as the world capitalist economy. In it the 'individual' is not in fact simply the national economy of each country (in which the universal is also manifested 105 in an infinite variety of partial cases and specifically national features of development), but primarily embraces various groupings of these countries, the most important aspects of whose economies are united by the most essential common features into major sub-systems of the world economy. In the present historical situation these sub-systems include, first of all, the developed capitalist countries and the developing countries, i.e. the former colonies and semi-colonies that have thrown off foreign domination and taken the road, within that economy, of anti-imperialist struggle for independent national development. It would be logical to begin our description of the changes in the postwar distribution of the non-socialist world's productive power with an analysis of the distribution of its main industries in the light of Lenin's law of the uneven economic development of capitalism.

Capitalist and Developing Countries

Colonialism put an indelible stamp on the structure of the geographical distribution of capitalist production. An evergrowing concentration of production capability in the comparatively few centres of imperialism and its slow growth in the overwhelming majority of the dependent agrarian countries and raw material producers was a major feature of the whole preceding colonial history of capitalist production. In disclosing one of the important aspects of the operation of the law of the unequal development of capitalism, Lenin remarked:

Capitalism develops unevenly and objective reality gives us highly developed capitalist nalions side by side with a number of economically slightly developed or totally undeveloped, nations.

The immense unevenness of the development of separate regions and countries on the one hand and their industries on the other is not obvious in the dynamics noted. This unevenness will be analysed from the standpoint of changes in the structure of the distribution of the productive forces by the principal geographical regions, groups of countries, and countries of the two major sub-systems of today's capitalist economy in the following sections of this chapter. Here we shall bring out the most general shifts to date in 500- Indices of capitalist and developing countries' production of GDP (1950=100) 450–400-350- 300- 250–200– 150- 100 &Lcirc; 1950 * Estimated *Sources*: calculated from the UN *Yearbook of National Accounts Statistics, Statistical Yearbook*, and *Monthly Bulletin of Statistics* for the relevant years the distribution of GDP between these two sub-systems. They give a very clear picture of how the economies of developing countries began to grow more rapidly than those

107 of the developed countries in the postwar period, especially in the mid-60s.

This has been happening on a considerable scale of late because of the distinct slowing of economic growth rates in the metropolises of capitalism during its world crisis of the mid-70s. Growth rates also slowed rather, in turn, in the developing countries under the impact of the crisis, but there they were nevertheless relatively higher on average than in capitalist countries, which could not help intensify the trend toward a faster growth of GDP in the former than in the latter shown in the graph.

This trend has become quite stable, and is showing itself in fact (though far from evenly) both in all their main industries and in the services and circulation sphere. Its action brings out a very important feature not only of the current shifts in the distribution of the capitalist world's productive forces associated with the break-up of the colonial system but also with possible prospective ones.

Study of this trend, however, without clarifying the specific character of its movement by industries, is clearly inadequate to characterise the patterns promoting probable changes in the future. From that point of view a comparative analysis of the structure and long-term dynamics of the growth of production in industries on the one hand and in the circulation and services sphere on the other, by the two groups of countries, is of great interest. The facts indicate that for all the diversity of their growth rates, the circulation and services sphere, etc., developed much faster in the agrarian countries (in contrast to the industrial ones) than the industrial sphere. As a result the proportion of the circulation and services sphere in the total gross production of developing countries rose roughly from 37.5 per cent in the early 50s to 44 per cent at the beginning of the 80s, while there was a slower development of this sphere on the whole in the industrialised countries, with the exception of periods of world cyclic crisis which, as a rule, bore most heavily on industrial production.

The trend toward faster growth of material production on the scale of the world capitalist economy noted above, however, operated mainly only before the world crisis of 1974–75, and stemmed basically from the features of the impact at that time of the scientific and technical revolution on the economies of the capitalist industrial centres. The structural 108 shifts emerging in the periphery developed in another direction; the circulation and services sphere expanded there much faster than the branches of material production. This difference of direction within a single system is an interlocking of aspects of the processes that are common to the whole system. In this case they include (in our view) not only the initial breaking down of age-old backwardness in sectors that form the infrastructure of the economies of former colonies and semi-colonies (railways, communications, etc.) but also the progressing impact of the scientific and technical revolution on the infrastructure of the world capitalist system as a whole. It is under its direct impact that the picture of the level and growth rates of production has begun to change, important changes are developing in the principal items of society's material and technical basis, and its sectoral proportions and the ratio between its productive and non-productive spheres are altering. The revolution is furthering advance of some sectors along the line of scientific and technical progress and a slowing of the development of others.

The steady deepening of the main contradictions of world capitalism in the 70s and early 80s, which was marked by a general slowing of economic growth within its limits, led to a sharp intensification of the instability and imbalance in the development of capitalist countries' social production. The stimulating effect of contemporary scientific and technical progress on the economy (above all on the leading industries) clearly weakened, and there were very substantial shifts in those years in that connection in the sectoral structure of capitalist and developing countries (and so in the whole system of their world economic relations) on a scale unknown in preceding decades.

In the summary analysis of the postwar structural changes leading to a raising of the weight of the circulation and services sphere in the gross production of capitalist and developing countries, the following facts primarily draw one's attention. Transport and communications stand out specially in both groups of countries. In 1960–81 alone their volume of production in the industrial centres increased by a little more than 140 per cent, and by nearly 250 per cent in the developing countries. As a result their weight in the GDP of each group converged, rising respectively from C.O to 6.5 per cent in the capitalist countries and from 109 rouglily 4.5 to 5.5 per cent in tho developing countries.

Although the proportion of transport and communications still remains comparatively low, its dynamics are very informative. It is in them that the tendency in today's capitalist world toward a certain smoothing out of the marked imbalances in the structural relations of the various sectors of the GDP between developed and developing countries is making itself felt.

This tendency began to become noticeable in the postwar period to one extent or another in almost all the principal sectors creating goods and services, including those that used to be decisive in characterising the underdevelopment of the colonial world's economic structure.

The corresponding indicator for developing countries was only a little more than 20 per cent. Subsequently this gap began gradually to close. Changes like that do not yet, of course, enable us to characterise the depth of the remaining backwardness of the structure of the economies of developing countries thoroughly and fully enough. The fact, for example, that the proportion of commerce in their GDPs was approximately the same as in the industrial centres of capitalism is by no means evidence that they have achieved as significant a degree of development in this respect. We also have to allow for the absolute volume of production of each sector in the one group and the other, the size of the population, and the level of labour productivity (which we shall touch on below). A comparative study of these structural shifts, however, enables us to size, up many of the important features of the postwar development and distribution of the productive forces on a very broad plane. As we develop the main theme of this chapter they will be consistently analysed more concretely and more exactly defined.

In this respect to describe the changing role of industry is of the greatest significance. The growth rates of industrial production in both groups of countries are higher than those of their social product; consequently the weight of industry in the GDP of industrially developed countries for the periods in a tendency to grow a little in the 50s and 60s, while it rose by 50 per cent in the peripheral countries. In the following decade, however, due to the two economic crises, the annual growth rates of industry did not rise in the first group, while the corresponding indicators were even lower in the aggregate growth of the GDP of the second group. As a result the proportion of industry in their gross production of goods and services roughly flattened out and was around one-third at the beginning of the 80s.

The longer-term trend arising in the postwar period toward a reduction of the imbalance in the structure of production between the two main sub-systems of the world capitalist economy had thus begun to show more pronouncedly in industry than in the circulation and services sphere.

In spite of the remaining wide gap in levels of industrial development, analysis of the current processes, that are making a further undermining of the colonial foundations of the capitalist system inevitable, allows us to suggest that this convergence will probably increase. One can foresee, in particular, that the difference in the weight of industry in the GDP of the two groups created in earlier stages of the genesis of world capitalism will be narrowed still further in coming years. In some industries, however, there may be a significant coming together in this century.

The same line is discernible to some extent in agriculture for all its extremely specific character and peculiarities. In both groups there has been a parallel and quite marked fall in recent decades in the weight of agrarian produce in their aggregate social product. The proportion of agriculture in the GDP of developed capitalist countries fell from 8.5 to 4 per cent, and in developing countries from 36 per cent to less than 20 per cent.

The indicators cited include the results of very intricate socio-economic processes and could undoubtedly be taken as the starting point for a special comparative study of the real shifts that are becoming noticeable in the agriculture of these groups. But they already provide an opportunity in their present form to make certain generalisations of a factual order needed for further analysis.

During the whole preceding colonial development of capitalism, a decisive part of the aggregate product of its periphery was, as we know, created in agriculture, which was a natural consequence of the underdevelopment of the social production of the colonies and semi-colonies. This underdevelopment was further deepened by the fact that the colonial authorities saddled them with the role of economic appendages of the metropolitan countries. Their agrarian production was forcibly oriented on supplying the centres of capitalism with produce at the expense of their own consumption, so that while agriculture was the main occupation of the bulk of their populations, hunger and poverty prevailed in them, which was one of the most pernicious consequences of the transformation of capitalism into a world economic system.

By the middle of the twentieth century the gap ih this sphere between the metropolitan countries and their agrarian, primary commodity producing appendages had attained colossal proportions. The weight of agriculture in the latter's GDP was then almost double that of industry and constituted nearly 90 per cent of the value of the production of all sectors of the circulation and services sphere put together. In the metropolitan countries, however, the weight of agriculture was correspondingly roughly a quarter of that of industry and 87 per cent less than the proportion of the services and circulation sphere. Since then there have been notable changes in the structural proportions, which is very important to allow for not only when estimating the shifts that have already occurred but also when calculating probable coming shifts in the distrubition of the world economy's productive forces.

As for the developing countries, it is worth noting that agriculture, whose share still considerably exceeded that of the industrial product at mid-century, has since finally and irrevocably fallen to second place behind industry. There are real grounds for predicting a further change in this respect. In our estimation the line of development noted may lead to the weight of industry in the GDP of developing countries being at least double that of agriculture by the end of the 80s. Their lag behind the industrialised countries, however, will still undoubtedly be very considerable then, since the industry/agriculture ratio in the latter was already estimated at around 8.5 : 1 in the 70s, and will probably not be less than 9 (10) : 1 in the 80s.

Comparison of the long-term trends of development of industries and direct services also leads to not uninteresting estimates of the changes occurring, or anticipated in the future, in the structure of the GDP of the two groups of countries. According to them the ratio between agrarian production and the services sphere, for instance, steadily changed in favour of services in the economic centres of capitalism. At the beginning of the 80s 14 times more commodities were produced in it (in terms of value) than in agriculture.

A similar process, but at a qualitatively different level, was observable in the last decades in the former colonial world as well. Even at the beginning of the postwar period the share of services in the aggregate product was around 60 per cent less than that of agriculture. Three decades later it had already risen much above the latter. To all intents and purposes, however, the services sphere will retain a clearly marked tendency everywhere in the foreseeable future toward priority development compared with agriculture, probably with accelerated growth rates where the backwardness of the relations of production previously maintained by colonialism has long been an obstacle to its establishment and development.

Comparison of the growth results of the other main spheres of material production (i.e. industry and the sectors constituting 'services') yields a diametrically opposite picture for the two groups. At the beginning of the postwar period the proportion of industrial production in the GDP of both of them (especially of the second group) was less than that of sectors belonging to the services sphere. But in recent decades the proportion of industry came noticeably closer to the aggregate share of the services sectors and later overtook them in both groups of countries. In other words, there was a smoothing out in this respect, too, of the acute imbalances in the structure of social production in both groups with the decay of the colonial system, while the one preserved its immense superiority over the other as regards level of economic development. We shall shortly demonstrate below how the structural proportions between the principal industries within each group gradually altered after the war, but already, on the basis of the facts cited, we can conclude that, in spite of the extremely broad variety of the trends of socio-economic development, the determinant features of the postwar dynamics of the growth of social production in the countries taken separately, and the structure of the distribution of production created by capitalism between its industrial centres and agrarian periphery, are beginning to undergo fairly important changes.

As the graphs indicate, there has been a long-term trend in literally all the principal sectors forming the gross product of the non-socialist world toward an enhancing of the role of the production of developing countries. In the period covered by their share in the industrial output of the capitalist economy has increased by roughly 75 per cent, in building and construction has doubled, in transport and communications has risen by a third, in services by twothirds, and in trade by more than a quarter. That does not, however, mean any kind of fundamental break in the longstanding extreme unevenness of the distribution of production capacities between industrially developed countries and the primary producers in the present day capitalist economy. In all its principal sectors, except agriculture, five to seven times as much output originated in the metropolitan countries at the end of the 70s as in the developing ones.

Even the indisputable fact that the developing countries surpassed the main economic centres in volume of farm production does not give us the right to consider there to have been any kind of essential turn in the distribution of the productive forces. Furthermore, it is agriculture that is still, as a rule, the most backward sector of their economies, which to a crucial extent determined the steady widening of the gulf between them and the developed countries in per capita gross production.

It is of paramount importance in this connection to analyse the regional distribution and long-term dynamics of population growth. This analysis, being a *sine qua non* of a comprehensive study of the distribution of the productive forces, also calls for an all-embracing systems approach. The next section of this chapter will be given over to it, which will enable us to present the socio-economic aspects of this problem more fully and clearly. But before we pass to a comparative description of the demographic structure of the capitalist economy by regions and groups of countries, we must dwell on at least one more matter of first-rate importance.

So far the objective has been to delineate the general contours of the dynamics, scale, and distribution of the gross domestic product in terms of capitalism's two principal groups of countries. At this very high degree of statistical abstraction we must not lose sight of the following point: each of the aggregate sectors considered in itself is an extremely complex system embracing at least several major lines of production which in turn are built up from ever more specialised groups of industries.

Within industrial production as a whole we primarily distinguish such very broad, large-scale sectors as manufacturing, generation of electricity, and mining, each of which can be pictured as a kind of sub-system within the aggregate, consisting of smaller sub-systems. Manufacturing, for example, is broken down into the steel, metalworking, chemical, textile, food, and other industries, which again include more specialised industries.

The logic of a macro-economic study is therefore the same in this case—from broad generalisations following from highly synthesised data to their elaboration and itemisation by analysing the main trends in the movement of the separate sub-systems in their interaction and subordination. It is through this approach that we get the opportunity to determine the specific features of the development of the part in the world economy concretely compared with, and integrally connected with, the principal channels of movement of the whole.

The most dynamic trend of industrial development in both groups was in the power industry, whose average growth rates were more than double those of manufacturing and four times as high as those of mining (which could not help raising its role). The proportion of the output of the power industry in all total industrial output, in comparable indicators, was around

3 per cent on the eve of World War II and is now more than 8 per cent. The power capability of the developing countries lias heen expanding particularly fast; in the 60s and 70s alone it rose by roughly 500 percent. In spite of that, however, their industrial production of power was still rather less than one-seventh that of the capitalist countries at the beginning of the 80s.

The difference between the two groups has remained even deeper, on the whole, in manufacturing, in which the weight of the developing countries has remained very low, even though it has increased slightly. At the start of the 80s it was roughly 14 per cent against 9 per cent at the beginning of the 60s. The weak development of manufacturing is the main reason for the preservation of many features of the old colonial structure in the distribution of the productive forces of present-day capitalism and the backwardness of all the other spheres of the economy involved in creating the former colonies' aggregate social product.

With a further deepening of systems analysis of the shifts in the distribution of the aggregate product between the two groups of countries in each of the sectors of production, we can distinguish major sub-systems, as in industry. In agriculture they include primarily the production of food on the one hand and of industrial raw materials on the other, which in turn consist of more specialised groups of industries (crop growing, animal husbandry, etc.).

When making a comparative study of the cardinal problems of the distribution of any industry, however, we must always ask what is the relation between the dynamics of production of the respective product in the different groups of countries, and the size of their populations and manpower.

The Principal Productive Force of Society

The growth of human resources in the much contracted postwar capitalist world economy has been unparalleled in scale. The population of all its countries increased by more than 1,350 million in 1950–80, which was more than double the corresponding figure for the first half of the century (620 million). As a result the total population of the capitalist and developing countries was nearly 3,000 million at the beginning of the 80s (as against only a little more than 1,600 million at the beginning of the 50s and 1,000 million at the beginning of the century). In other words, human 119 resources increased by 80 per cent between 1950 and 1980 within the non-socialist world, and by roughly 190 per cent since the beginning of the century.

These considerable changes were determined to a crucial extent by the 'demographic explosion', whose causes are linked with the break-up of the colonial system and advances in health protection. These points have been widely discussed in Soviet socio-economic literature, so we shall touch here only on those consequences of population growth that characterise its distribution in the principal regions of the capitalist world and are essential for understanding the shifts in the distribution of manpower among them, and consequently of the long-term trends in the development of their basic productive force.

Such an immense disproportion has not, by any means, been typical of the whole period of the development of capitalism's world system. Estimates based on UN demographic statistics indicate that even at the beginning of this century the growth rate of the metropolitan countries was patently higher than in the colonies and dependent countries of Asia, Africa, and Latin America, where 120 epidemics raged and hunger and high infant mortality prevailed. The expectation of life in the latter was about half that in the metropolitan countries. Since the war there has been a marked exacerbation of the demographic situation in the capitalist %orld. The rates of increase began to rise in the former colonial world. In 1938–50 they averaged 1.4 per cent a year, in the 50s 2.3 per cent, and in the 60s and 70s 2.6 per cent. In the developed capitalist countries the rates were also rather higher in the first postwar decades, but again began to fall and in the 70s were around 0.85 per cent a year, i.e. about the average for 1900–1938.

The most populous centre of world capitalism has always been and still is Western Europe. At the middle of the century rather more than half of the total population of capitalist countries lived there, but the growth rate in most West European countries has been very low in recent decades compared with other geographical regions, while emigration has been very high, which has deepened the trend already long apparent toward a reduction of the weight of Western Europe's population in both the capitalist countries as a group and the capitalist world as a whole.

The population of the North American area increased much faster than Western Europe's after the war. In the period 1950–81 its annual average growth rate was 1.30 per cent, which led to a certain convergence of the absolute size of the populations of the two main centres of capitalism. While the ratio between them was 1.7 : 1 in favour of Western Europe in the early 50s, it was 1.4 : 1 at the end of the 70s.

p One must remember, however, that the population of North America was increased in the first half of the century, and especially as a result of World War II, by a flow of immigrants from outside, including Western Europe, in addition to a rate of natural growth higher than in other developed capitalist countries. At the same time there was a high population growth rate in other capitalist countries, above all in the biggest of them, Japan, although it was also much lower than in the developing countries. A consequence of all these factors was a considerable decline in the weight of the industrial centres in the total population of today's capitalist world.

The population of Africa (not counting South Africa) rose by 210 million, or more than doubled, in the same period, and its average annual growth rate began to exceed that of Asia in the 70s. These rates were not, however, limiting ones. Latin America set records of sorts, its population increasing by 120 per cent by the end of the 70s compared with the early 50s, with a mean annual growth rate of 2.75 per cent.

These major shifts in the ratio of population among the various groups of countries convincingly emphasise the need to make careful allowance for the demographic factors when studying the long-term development trends of the productive forces in the postwar capitalist economy. And they are of special significance for a sectoral analysis of the results of the main regions' economic development.

At the same time these factors are extremely important for forecasting growth of the productive forces in any one economic system and on a global scale. The Soviet sociologist Arab-Ogly has justifiably commented that the tempestuous growth of population is one of the most important historical phenomena of our time as regards its remote consequences. It is facing both individual countries and mankind as a whole with a number of exceptionally complicated though in principle resolvable economic, social, and political problems. That is why there is an urgent need for substantiated and as exact as possible forecasts down to the year 2000. The compilation of such forecasts has now acquired universally acknowledged theoretical and practical significance. Demographic processes as a rule have immense inertia, which also makes it possible, strictly speaking, to make more or less substantiated forecasts of future populations.

The figures quoted above allow us to slate that the longterm trends in the area of population are more amenable to projection into the future than those of production, given the instability of development integrally inherent in the capitalist system, cyclic crises, feverish booms and slumps in market conditions, and so on. In spite of the marked rise in tempos in various countries and years, the dynamics of population growth is limited in the principal regions by quite definite parameters, which provides the necessary premises for a more or less long-term estimate of the future distribution of manpower among them.

A real effort has been made in UN demographic publications to tie up population growth trends somehow with certain outlooks for the subsequent course of development of production. One of its surveys remarked, for instance:

> The future size, structure and distribution of population are essential for any plan that involves food, housing, employment' education, health or other public services. This is also an era of increasing awareness: of man's unprecedented growth, of the interaction between this growth and the environment, and of the possible implications for the future. Investigations into these complex relationships have added to the demand for population projections.

For all the substantial differences in estimates of the future development of the international demographic situation, the majority of them quite validly, in our view, assume that one can hardly expect any extraordinary changes in growth rates in the really foreseeable term either in the capitalist or the developing countries compared with the postwar period, though some demographers' forecasts assume a certain lowering of these rates in coming years.

In any case, however, there are adequate grounds for staling that the number of inhabitants of all the countries of the non-socialist world will increase by at least 50 per cent in the last two decades of the century (16 to 20 per cent in the first group and 56 to 60 per cent in the second). In absolute terms this means an increase in the population of the industrialised countries of roughly 120 to 150 million, and in the developing countries of 1,200 to 1,300 million.

These estimates, like all similar generalising forecasts, are approximate, but in themselves they are very important not simply for understanding the outlooks for the development of population but even more for clarifying their interconnections with the possible trends of the future growth and distribution of the productive forces in the countries at present forming the world capilalisl system. The postwar period has been marked by a further increase in the unevenness of the distribution of the main productive force in the developing world. Most of its inhabitants, and so of its manpower, has long been concentrated, as we know, in Asia, whose weight in the total population of the capitalist world has been steadily rising in postwar decades. Analysis of current trends gives grounds for sugg-esting that the Asiatic population will be around half in the mid-80s, and more than half in Ihe 90?, of the total population of the capitalist world within its present boundaries. But the majority of the liberated countries of Asia (in which more than 2,000 million people will probably be living at the end of the century, i.e. three times as many as in the middle of the century) are now the most backward areas, economically speaking, of the developing world. The population of Africa and Latin America is increasing even faster.

These expected shifts arc undoubtedly fraught with serious socio-economic consequences for the fate of capitalism in those areas. In coming decades jobs will have to be created in them for hundreds and hundreds of millions of people, but their agriculture is already unable to absorb any considerable part of the flood of new labour power. The objectively operating laws of agricultural production, moreover, are leading to surplus manpower being pushed out into the towns which, given capitalist relations of production, inevitably means a growth of poverty and proletarianisation of broad masses of the peasantry. That is one of the important distinguishing features of capitalism compared with the exploiter formations that preceded it. The law of faster growth of the urban population under the capitalist mode of production, Lenin stressed, 'does not and cannot operate otherwise than through the disintegration of the peasantry into a bourgeoisie and a proletariat'.

When that conclusion was drawn there was still no socialist economic system. Experience of its development has shown that for it advance of society's productive forces is inseparably linked with priority growth of the urban population, but it also indicates that the new mode of production, unlike the capitalist, makes it possible to avoid the extremely grave consequences of urbanisation for the broad masses objectively inevitable for all capitalist countries.

Since the war urbanisation has intensified to an unusual degree under the impact of the shifts in the structure of social production considered above. It has begun to have an ever more active effect on the character of the distribution of manpower in the world capitalist economy. In the 50s, 60s, and 70s the urban population of non-socialist countries more than doubled, so that the ratio of urban and rural population altered, with socio-economic and political consequences of no little importance.

At the end of the 70s more than 40 per cent of the capitalist world's total population was already living in towns, while less than a third had done so in the early 50s. In the industrial countries this proportion rose to 75 per cent, and in the developing countries to a third, with fewer and fewer of the rural inhabitants available for work being directly involved in agrarian production. At the same time an increasingly large amount of manpower is becoming concentrated in the towns, a trend that has become stable not only in industrially developed capitalist countries but also in the overwhelming majority of developing countries.

There are, of course, considerable regional differences in this respect in both groups, largely associated with features of their preceding historical development. In the industrial centres the highest proportion of urban population is in North America, particularly in the United States, where more than 90 per cent of the total population was living in towns at the end of the 70s. There is an even more significant spread in the weight of the urban population in developing countries, from 15 or 20 per cent in Asia to 60 or 65 per cent in Latin America.

One of the most characteristic trends in recent urbanisation is the rapid increase in the role of former colonial regions in the total urban population of the capitalist world in both absolute terms and relatively. At the beginning of the period reviewed the bulk of the urban population lived in industrial countries, while at the end already more than half of the townsmen lived in the agrarian and primary producing countries. The size of this population grew roughly twice as fast in the latter as in the former, and analysis of the long-term trend allows us to suggest that more than half of the population of the contemporary capitalist will be townsmen by the end of the century. At least twothirds of them will be living in the former colonies and semi-colonies of Asia, Africa, and Latin America.

A not unimportant feature of the long-term shifts in the distribution of population is its growing concentration in very big cities. The number of cities with more than a million inhabitants increased in the 50s and 60s alone by more than 50 per cent, while the number of their inhabitants more than doubled. Big cities have been growing much faster, moreover, in developing countries. Most of the biggest cities of the contemporary capitalist, world are now already located in developing countries. The chaos of capitalist urbanisation is leading to the formation of broad urbanised zones with multimillion populations in both groups of countries, which are drawing an ever bigger part of the population of adjoining towns and rural areas into their orbit, thereby intensifying the unevenness inherent in capitalism and anarchy in the distribution of society's main productive force.

Substantial shifts are also taking place in the geographical distribution of the rural population of the non-socialist world, which has grown in recent decades exclusively in the agrarian and primary commodity producing countries. In the industrially developed countries there has been an absolute fall in its numbers, while it has increased by more than two-thirds in the developing countries as a whole, and exceeded 1,600 million at the end of the 70s.

In the principal economic centres the complete ousting of the petty producer from agriculture was completed, or almost so, in the postwar decades. Engels pointed out the objective inevitability of this in his time, remarking that the development of the capitalist form of production has cut the lifestrings of small production in agriculture; small production is irretrievably going to rack and ruin.

In the majority of emancipated countries that have taken the road of capitalist agrarian relations, this process is developing with increasing force, but attempts to force its pace are still complicated by the low level of development of the other spheres of their economies, above all of industry.

The proportion of the industrial countries in the total rural population consequently fell by half in the 50s through 70s, while it rose from less than 75 per cent in the developing countries to 90 per cent. As a result there was a considerable extension of agrarian overpopulation in the latter. This enormous overpopulation, which was long a heavy burden on their economies, still remains one of the most disastrous social 'side-effects' accompanying their economic growth under capitalism.

An increasing amount of world capitalism's rural population is thus concentrated in developing countries, above all in Asia and Africa, where the rapidly growing towns are unable to absorb the villages' immense surplus of manpower. At the beginning of the 80s around two-thirds of this rural population lived in Asia and more than 20 per cent in Africa. The only major region where its proportion has begun to fall is Latin America, and there too, in fact, it still maintains its tendency to grow in absolute terms (increasing by more than 40 per cent in the period under review). In the coming decades the marked imbalance will apparently not be smoothed out, but will become rather more acute. In fact, all in any way valid long-term demographic forecasts envisage a progressive reduction of the rural population in capitalist countries and an inevitable growth of it in developing ones by at least a third by the end of the century.

The absence of sufficiently reliable estimates of the agrarian overpopulation in developing countries is a factor making it extremely difficult to determine exactly the scale and dynamics of employment in the whole world capitalist economy. The inadequacy, in turn, of the data on full or partial unemployment in industry and other spheres of production in most developing countries, and in some capitalist ones, is also an obstacle; so, too, are the constant upswings and recessions in employment through the effect of marked cyclic and non-cyclic fluctuations of the capitalist economy.

Still, however, there arc composite indicators in the international statistics that enahle us to estimate, with the necessary degree of comparability, the growth trends in the long term of that part oi' the population that could be involved in the reproduction process. This type of indicator primarily includes estimates of the scale of growth of the gainfully employed population, analysis of which helps us arrive at the characteristics of the development trends of productivity for the different groups of countries.

The imbalance in the postwar growth of the able-bodied population between the two groups of the capitalist economy has increased immensely. Over the twenty years of the 60s and 70s, for example, their numbers increased twice as slowly in the developed capitalist countries as in the developing ones. Consequently there was a tendency for the proportion of the former in the total gainfully employed population of the capitalist world to decline.

From the fact that the decisive mass of the population that will begin to work at that time in both groups of countries already exists, we can assume that around three–quarters of the gainfully employed population of today's nonsocialist world will be concentrated in the now developing countries of Asia, Africa, and Latin America in the mid-80s, and around i'our-iifths by the end of the century. But this labour force will still obviously be relatively unskilled in the main while the most skilled part will be concentrated as before in the economically developed centres.

In that connection the problem of the migration of various categories of workers between countries with different levels of development will be aggravated. In many industrial countries there has been a fast growing shortage since the war of national personnel in the mass trades that do not call for any degree of professional skill. The shortage began to be covered more and more by importing labour from economically less developed countries. In the 70s this shortage constituted around ten or twelve million, and may (according to some estimates) increase several times over again by the end of the century.

As for the developing countries the inherent tasks of their socio-economic development will steadily lead, in the first place, to an increase in their need for well-trained, skilled personnel, a need that cannot be wholly met simply from local resources. The possibilities for attracting specialists from outside, however, are greatly limited by the increasing need for such personnel in the developed countries themselves, and also by the comparatively high standard of their pay there. It is not fortuitous that a mounting portion of the West's funds for 'technical aid' are spent on the specialists sent to developing countries, and on training students.

At the same time, monopoly capital, utilising the laws of capitalist competition, is everywhere broadening the 'import' from developing countries of the most talented and capable part of their intelligentsia and skilled specialists. The 'brain drain' from Asia, Africa, and Latin America has become a marked feature of the international migration of labour. The spontaneous flow of surplus labour between separate developing countries has also become unparalleled in scale in recent decades.

All these objective 'costs' of the internationalisalion of social production are making world capitalism's demographic problems extremely complicated and increasing their tension. At the present level of development of mankind's productive forces the registering, regulation, and planning of manpower are already becoming more and more urgent on both the national and the international level, but capitalist society, rent by antagonistic contradictions, is unable in practice to face up to dealing with them. They can only be coped with by a more rationally structured progressive social system, a thought, that Engels expressed as follows:

If it should become necessary for communist society to regulate the production of men, just as it will have already regulated the production of things, then it, and it alone, will bo able to do this without difficulties.

In most areas of the capitalist world there has been a tendency since the war toward a certain reduction of the proportion of the gainfully employed population in the total population. The operation of this trend is linked in the main with the considerable changes in the age structure of the population, especially in the developing countries, where there has been a marked extension of the age brackets of people who have not yet reached maturity. In many capitalist countries, above in Western Europe, another factor underlies the lowering of the proportion of the ablebodied, namely the rise in the expectation of life and increase in the age bracket of elderly people of pensionable age. Analysis of the age factors leads to the conclusion that an increasing part of the most active age brackets will probably be concentrated in coming years in the developing countries, from which there follow very complex demographic, and consequently socio-economic, consequences for further shifts in the distribution of the main productive force within the world capitalist economy.

The indices of the dynamics and scale of the distribution of manpower in the two groups of countries will become even more significant when they are compiled from the data cited above on growth of the aggregate product. This comparison makes it possible to bring out the long-term trends in growth of the social productivity of labour calculated on the labour force potentially existing in the non-socialist world, i.e. on the size of the gainfully employed population. This productivity can be expressed most generally as the difference between the mean annual growth rates of the GDP in every macro-economic system (or its sub-systems) and the corresponding indicators of the growth of manpower.

Because of their generalised character the results arrived at are very approximate and require further elaboration and concretisation both for the groups of countries and their leading industries and for the various periods. It follows from them that, with all sorts of fluctuations of production and employment in the two groups of countries, the mean annual growth rate of potential productivity of social labour, calculated from the dynamics of the gainfully employed population over the whole period considered was around 3.1 per cent in the centres of capitalism and 3.2 per cent in the developing countries. This relation was extremely unstable, however, and was by no means typical for all the postwar decades.

In the first years after the war the considerably accelerated economic development in the main agrarian economic regions, and the comparatively low rates of manpower growth preserved in them from the colonial epoch, led to a tendency toward accelerated growth of their extremely low productivity of social labour, reaching the average indices of the main centres of capitalism in the 50s. In the 60s, however, there was again a revival of capitalism's inherent line toward widening of the gap in this respect belwcen its centres and periphery, which was furthered by the industrial countries' active taking up in those years of the advances of the scientific and technical revolution, though within the system as a whole insuperable inner contradictions fraught with new economic upheavals continued to accumulate.

The extremely acute crisis processes that developed in the rnid-70s in the world capitalist economy were marked by an, unprecedented weakening of the economic effectiveness of exploiting the advances of this revolution in practice in all Ihe industrially developed countries. The substantial slowing of growth and later the absolute fall in production of their aggregate GDP could not help being reflected as well in the productivity of social labour.

The major postwar economic crisis of the mid-70s laid bare the growing instability of the world capitalist system and the extremely contradictory character of the effect of modern scientific and technical progress on it as regards growth of society's productive forces. It would hardly be proper, however, to say that the objective patterns leading to the creation of such a vast imbalance in the levels of the social productivity of labour between the two groups of countries at the different poles of this system will begin to lose their former significance in coming years. In all probability this imbalance and steep fluctuations of its dynamics will remain essential features of the world capitalist economy's further development.

The processes mentioned indicate simply certain very general resultant trends in the postwar shifts in the economies of capitalist and developing countries. Within each of these groups there is in turn a very wide range in the levels, dynamics, and conditions of the growth of social productivity. Above all we must draw attention to the considerable strengthening of unevenness in the distribution of the productive forces in the developed capitalist countries in recent decades.

The Industrial Centres of Capitalism

The decisive part of (lie economic potential of capitalism is still, in spite of the very considerable changes in the distribution of its productive forces after World War [I, concentrated in two principal regions–Western Europe and North America. This is due to the whole preceding course of their historical development, especially in the monopoly stage. At the beginning of the 1080s more than 70 per cent of the aggregate product of world capitalism was created in these regions, or nearly seven-eighths of the gross production of its industrial centres.

As a result of World War II there was an abrupt redistribution of forces between the principal countries of monopoly capitalism and their opposing groupings of that time, but the total share of the production of North America and Western Europe within the shrunken boundaries of the imperialist system after the war not only was not reduced but even rose a little. At the turn of the 40s and 50s around 94 per cent of the goods and services of all the countries of this system were produced in them.

The postwar disorganisation in the economies of the overwhelming majority of West European countries and Japan favoured establishment of the domination of the United States in the capitalist world, production capacities being concentrated in it that considerably exceeded those of all the other capitalist countries taken together. Major sectors of the economy of Canada, where around 6 per cent of the GDP of the North American region was created in 1950, were moreover under full American control.

The economic expansion of U.S. monopolies had acquired a broad scale in other developed capitalist countries as well, since the latter's monopoly groupings were at best only in a condition to claim the position of junior partners of their American rivals. U.S. military and political superiority in the non-socialist world was then unchallenged. Priority of the dollar had also been established in the international monetary system. It would seem, at first glance, that there was no force that could oppose the gigantic growth (as a result of the war) of the expansionist strivings of the American monopolies.

And still the attempts to consolidate hegemony of one country's monopoly capital in the imperialist system further, or to maintain it for a long time, proved flimsy. The objective processes governed by operation of the law of the unevenness of capitalist countries' economic and political development, were characterised by new shifts in the distribution of the productive forces within modern capitalism.

Although the GDP doubled in North America in the 50s arid 60s its proportion in the world aggregate fell by around 5 per cent. The volume of production in Western Europe rose by 160 per cent in the same period and by nearly 360 per cent in the other capitalist countries, mainly because of Japan. Consequently, as is seen from Fig. 8, there was a corresponding growth of the proportion of these two last groups, and the gap between the GDP of North America and Western Europe was patently reduced.

The operation of this trend during the cyclic development of modern capitalism, it is true, like the processes in general considered here, has not been displayed evenly or stably. After the world crisis of 1974–75, for instance, the gap between North America and Western Europe widened a little, but at the turn of the 70s and 80s, because of a substantial fall in GDP growth rates in the USA and Canada, it again narrowed. Since the late 60s Western Europe's share in the total production of capitalism's industrial centres has in turn begun to show a tendency to fall (because of the speeding up of the economic growth rates of other developed capitalist countries).

This has all led to the total weight of North America and Western Europe in the total GDP of all capitalist countries beginning gradually to decrease in the past 15 or 20 years, for all the considerable fluctuations in their development rates. By the beginning of the 80s it was 8 per cent lower than at the beginning of the 50s, and 5 to 6 per cent lower than the 60s. It is unlikely that one need consider this line of development a temporary one.

In the capitalist countries not located in the West European and North American centres, the principal production capacities are concentrated in Japan. It is the rapid postwar development of the Japanese economy that has governed the substantial growth of the role of this group of countries in the world capitalist economy. At the end of the period reviewed their share in the aggregate production of goods and services of capitalism's industrial centres was more than 14 per cent, as against roughly 8.5 per cent at the very beginning of the 60s, and 6.5 per cent at the turn of the 40s and 50s. It is not fortuitous that the monopoly circles of the USA and Western Europe are increasingly alarmed by the possibility that the Japanese 'centre of force' will take the lead in the world capitalist economy by the end of this century.

The second area of significance among these 'other countries' is Oceania, which includes Australia and New Zealand. But they lag markedly behind Japan and the other leading capitalist countries in both the scale and the rates of their development of production. At the end of the period under review their production was only a fifth of Japan's, while their GDP had been more than half that of Japan's at the beginning of the 50s. Oceania, and in particular Australia, has great, still unutilised potential resources, which allows us to assume a rise in the development rates and role of this region in the coming period. We must also remember, moreover, that the countries of Oceania are being more and more distinctly drawn into the orbit of economic influence of the Japanese 'centre of force'.

The rivalry of the monopolies in the struggle to divide and re-divide spheres of political domination and economic influence can only be studied in motion. At the same time it would be wrong to reduce this struggle simply to an analysis of imperialist contradictions within the limits of separate countries or even regions. The whole system of capitalism is becoming its field. As Lenin wrote: Monopolist capitalist associations, cartels, syndicates and trusts first divided the home market among themselves and obtained more or less complete possession of the industry of their own country. But under capitalism the home market is inevitably bound up with the foreign market— As the export of capital 141 increased, and as the foreign and colonial connection;, and 'spheres of influence' of the big monopolist associations expanded in all ways, things 'naturally' gravitated towards an international agreement among these associations, and towards the formation of international cartels.

In those conditions a need arose for a comprehensive analysis of the process as a whole.

A new level of the international concentration of capital and production has been reached in recent decades, incomparably higher than at the beginning of the century. But whatever the degree, forms, and methods of concentration, Lenin's methodology of studying shifts in the structure and distribution of production among the principal imperialist powers remains of first-rate value for estimating the important development trends of the system of international capitalist division of labour and inter-imperialist contradictions. These shifts stem from processes developing within the national economies and characterising the position of each country in the world economy to a decisive extent.

The tendencies noted above toward growth of production in the industrially developed regions have in fact been determined in the postwar period, as throughout the preceding history of monopoly capitalism, by a handful of powers, whose composition has essentially remained unchanged, limited to the Bix Six (the USA, Japan, and the four West European powers, Great Britain, West Germany, France, and Italy).

Although there have been changes in the balance of power of the Six that have been of considerable importance in their long-term world economic consequences, more than four-fifths of the GDP of all capitalist countries was produced in them at the end of the 70s, a fact that explains one of the deep-lying causes of the maintenance of their dominant position in the decision of the world economic and political problems of modern capitalism. The tendency noted in postwar years toward a gradual reduction of the weight of the 'great' powers in the aggregate production of capitalist countries has primarily been associated with the very substantial fall in the postwar share of the USA. The objective conditions were thus brought about for an inevitahle gradual growth of instahility and inner contradictions in imperialism's system of world economic ties that arose soon after the war under the aegis of the United States.

The quite slow economic growth of Great Britain, the country that was the USA's closest junior partner in the building of this system, in turn helped weaken and undermine it. The share of Great Britain, once the mightiest power in the capitalist world, in the aggregate GDP of today's industrial centres of imperialism has fallen by nearly 40 per cent. At the same time there has been a relative consolidation of the positions of the other members of the Six (West Germany, France, Italy, and especially Japan), though with substantial differences.

As a result the distribution of the production capabilities of the main capitalist powers had acquired qualitatively new features at the beginning of the 80s, different from both the prewar and the first postwar years. The powers that were defeated in World War II, whose economies suffered a smaller burden of militarisation immediately afterward because of certain circumstances, strengthened their position in the capitalist world economy at the expense of the powers (chiefly the USA and Great Britain) that headed the imperialist camp after the war ;md were striving to take on responsibility for its fate. The sinking of Great Britain from second place among the Six to next-to-last, and the rise of Japan from last to second place, as regards GDP, are very indicative. West Germany perceptibly strengthened its influence, and only France, of the West European victor-powers, managed to consolidate its position a bit as regards its weight in the aggregate production of the principal capitalist countries.

A very peculiar situation built up during their uneven economic development in the 70s and early 80s because of the extraordinary deepening of the whole set of world capitalism's contradictions. Its peculiarity consisted primarily in the process' having taken place, as we showed above, on the background of a general and relatively protracted slowing down of the economic growth rates of all capitalist countries, which showed particularly clearly after the 1974–75 economic crisis.

It has therefore been a matter, of late, of the maximum prevention of a further crisis decline in rates of production, in the course of the bitter rivalry of the imperialist monopolies and powers on their home and the world markets, rather than of raising them. In that connection the shifts in the ratios of the industrial potentials of the leading industrial powers considered above show only the relative changes in their positions compared with other powers. The continuing rise in Japan's share of the gross product of the Big Seven, for instance, is essentially due to the fact that its rates of general economic development, although down 60 per cent in the 70s on the previous decade, all the same remained markedly higher than those of its main rivals. The shifts in the distribution of the productive forces noted above cannot, however, diminish the fact that the United States is still the leader among the capitalist powers because of its predominant economic might. Its share in the aggregate product of the Six, it is true, has shrunk, but more 144 goods and services were still produced in that country at the end of our study period than in all its principal imperialist rivals. The United States had a GDP four times that of Japan at the beginning of the 80s and almost double that of the four West European countries taken together. Some of the GDP of the other capitalist countries (apparently at least 6 or 7 per cent on the average), it must be noted, moreover, was created then either under the direct control of American transnational corporations, or with their direct involvement.

The loading position of the USA was markedly strengthened in the monopoly stage. In 1913 the USA produced almost as much output as Great Britain, France, Germany, and Italy taken together. In the early interwar period (1920) (according to the Soviet economists Bolotin and Kudrov) the volume of U.S. production was around half the GDP of the industrial centres of capitalism (within their contemporaneous frontiers), was about equal to the total GDP of all the countries of Western Europe, and was 25 per cent greater than that of the four main ones.

p From the end of the 20s the U.S.'s role in the world economy began to decline noticeably, and it felt the weight of the economic crisis of 1929–33 more than the other capitalist powers. On the eve of World War II the U.S. share in the aggregate production of capitalism's industrial centres had fallen to 41.7 per cent, but was still 230 per cent that of Great Britain (12.5 per cent), 340 per cent that of West Germany (9.(5 per cent), 380 per cent that of France (8.7 per cent), and six times as great as Japan's (7.0 per cent).

Generally speaking, long-term analysis of the constant fluctuations in the relation of the main powers' production capacities suggests that the position of the USA in the GDP of imperialism's centres was substantially weakened at the beginning of the 80s compared with the levels reached after the two world wars. Furthermore, their weight seldom fell so low during the whole preceding history of monopoly capitalism, yet, being still much superior to their main rivals in economic power, area, natural resources, and population, they still retain a leading position in the world economy.

It would obviously be insufficient to explain the changes mentioned just by transit factors stemming from the specific conditions of the development of production in these countries, taken separately, for one year or another. Changes like these on the long-term plane are ultimately determined by the deep-seated patterns of monopoly capitalism brought out by Lenin at the beginning of this century, patterns that include tendencies to a levelling of the development of its business centres.

This trend, which reveals an important feature of the intensification of the uneven development ol Ihe capitalist powers, like other objective patterns at the stage of imperialism was not and could not he stable. On the one hand, its effect has been repeatedly disrupted and broken by the military and political, and cyclic, upheavals of the world capitalist economy. On the other hand, it has been intensiiied and stimulated by the internationalization of capital, which in turn has promoted a concentration and specialisation of production on an international scale, mainly in periods of relatively peaceful development.

The processes leading to this levelling have been very distinctly manifested during the mounting impact of the scientific and technical revolution on the productive forces at a time when world socialism has significantly limited monopoly capital's possibilities of unleashing global destructive wars and resolving inter-imperialist contradictions by force of arms. Graphic evidence of that is provided by comparative indicators of the total GDP per capita in the countries where the main economic potential of modern capitalism is concentrated.

It will be clear from the facts cited that the unevenness leading to gradual closing of the gap between the development levels of advanced capitalist countries is being converted into an integral world economic problem of monopoly capitalism. With time, as these levels converged, it has been becoming more and more pressingly important, in fact, for the many phenomena determining both the current and the long-term development trends of capitalism's whole system, of international relations to be investigated. This tendency makes it possible, as well, to determine more exactly certain long-term consequences of the various groupings' continuous competition, and the chances, limits, and prospects of their temporary alliances and blocks in the postwar capitalist world.

It has begun to appear probable that some of the main imperialist countries will succeed in approaching the USA in such important economic indicators as GDP per capita even before the end of this century, and in individual cases overtaking them. It must be remembered, when analysing the world economic consequences of the shifts brought out in Fig. 10, that the U.S. growth rates of per capita GDP were higher in our study period than in the interwar years, arid higher on the whole than in the first half of the century. At the end of the 70s they were almost 80 per cent higher than in 1950 (in 1970 prices), but the dynamics of their growth was clearly slower than in the USA's main imperialist rivals, Great Britain excepted.

Consequently there was a marked redistribution of forces within this group of countries, and at the same time a certain convergence of levelsof development. The Japanese economy, for instance, which still lagged around 45 per cent behind Great Britain in this respect had surpassed the level of the latter by 50 per cent two decades later. Italy, which had been in bottom place in the Big Seven, had come very close to Great Britain, although, like Britain and Japan, it had not reached the average European level of per capita GDP at the end of the period analysed. At the same time France and even more West

Germany were considerably above this level, the latter proving to the even a little above the USA at the beginning of the 80s.

These facts by no means exhaust analysis of the extreme contradictions and trends spontaneously operating in the present capitalist world toward intensification of the uneven economic development of separate regions and countries, but they do make it possible to discern the general contours of a very intricate process, namely, the shifts in the'secloral structure of the distribution of their production, and so to bring out several characteristic features of this mounting unevenness. If we estimate these shifts in the light of the indicators of the dynamics of the main industries that contribute to the aggregate product of all the developed capitalist countries, and primarily of North America and Western Europe, then calculations based on UN statistics make it possible to draw the following conclusions, highly important for our further investigation.

The sphere of material production in North America (industry, agriculture, building) developed as a whole more slowly (with a very broad range of long-term and annual fluctuations of growth rates in the industries of each country taken separately) than in the countries of Western Europe. There had been even a more considerable slowing of North America's growth rates compared with the capitalist countries of other geographical areas, in which Japan's unusually rapid economic progress played a paramount role in the first postwar decades. We estimate that the volume of production in this sphere increased on the whole by 160 per cent in the first group during the 50s, 60s, and 70s, by 240 per cent in the second group, and by more than 500 per cent in the last. The gap between these regions, moreover, was not evenly closing in the various industries. It was closing mainly by industry and building. In agriculture, in which production developed at low rates almost everywhere after the war, North America was a little behind Western Europe, and later fell obviously behind the other capitalist countries.

As regards the sectoral dynamics of the sphere of circulation and services in the main regions, large-scale changes also developed, which can be attributed initially to several resultant indicators. The total volume of the product of the sectors of non-material production trebled in Western Europe over the same period, increased by 180 per cent in North America and nearly eightfold in the other countries where it had been historically at a relatively low level. Here, too, the levelling was accompanied with an intensifying of the unevenness of development.

Because of their high degree of statistical generalisation these figures cannot claim to be absolutely accurate. Still, in our view, they reflect the long-term processes actually taking place after the war with sufficient reliability. They can also be used as an approach to forecasting these processes, as they indicate the most probable trends in the movement of the interconnected tendencies both toward a heightening of unevenness and to a levelling up of the sectoral structure of capitalist countries' GDP.

It must also be recalled that as regards total volume of industrial output (in fixed 1970 prices) the West European region came noticeably closer to the North American, although the gap between them remained "quite considerable in per capita terms, being around 40 per cent in contrast to the nearly two-thirds of the early 50s.

The growth rates of the economic capability of Western Europe and other capitalist countries are inseparable from the development of construction in them. This has been the second industry (afler agriculture) in which Western Europe has exceeded North America in absolute terms since the war, though it still lags behind in per capita terms. At the same time the other capitalist countries have come fully up to the European level; whereas the contribution of building to the GDP in them used to be traditionally in last place, now substantially exceeds agriculture. At the end of the 70s the ratio of building and agriculture in the aggregate social product of all main groups of capitalist countries is estimated at 1.3:1 in favour of the former.

The international political consequences of the structural shifts that have taken place in the distribution of farm production in the capitalist centres are no less important than the economic ones. They are not as impressive and marked, it is true, as in industry and building, largely because of the historical features of the way the uneven distribution of the productive forces came about in this industry.

The West European area, which long surpassed the North American in the absolute scale of farm production (except during the world wars), has steadily fallen behind the USA and Canada in per capita production, especially of food. During the first postwar years this gap began to be rapidly closed, and in the 60s, when farming was transferred onto a basis of machine industry, West European per capita production was already around 60 percent of the North American level. The subsequent convergence was not, however, clearly expressed, since the production of certain food and fodder crops accelerated in the USA and Canada and was more and more oriented on the expanding demand of the foreign market. In some years the tendency to close the gap iii general disappeared, and was reversed.

The close dependence of agrarian production on natural conditions, plus the continuous fluctuation of market conditions, makes it difficult to distinguish stable trends and growth prospects in separate regions and countries over comparatively short periods. Still, the regional lines of the longterm development of agriculture noted above permit us to 153 assume that the tendency to level up will most likely continue given the potentially increasing shortage of certain types of agricultural raw material and food in the present-day capitalist world economy, above all in its former colonial periphery. The decline in the importance of agrarian production in the economies of most developed capitalist countries, it must be emphasised, will probably reach a certain limit, which may be characterised by the weight of agriculture in the GDP of the United States.

Everywhere there is, for example, a steady decline in the proportion of agriculture in the sphere of material production. In the period under review it fell from 14 to 8 per cent in North America, from 20 to 10 per cent in Western Europe, and from 30 or 31 per cent to 11 per cent in the other areas. At the same time there was a gradual raising of the corresponding proportion of industry almost everywhere in the main centres of capitalism, mainly through the leading heavy industries.

Among the objective processes operating synchronously in the sphere of non-material production the long-term trend toward a raising of the role of services has become of paramount importance since the war. The trend has developed unevenly in the different countries and at various stages, which has promoted the redistribution of the productive forces in the 'services' of the main economic centres already referred to. North America has maintained its leading position in this sector to a greater extent than in others. At the end of the 70s, for instance, around 56 to 57 per cent of the whole output of the services sphere of the developed capitalist countries was still produced in the North America, and it was, moreover, the most dynamic sphere of the postwar economies of the USA and Canada.

The superiority of the North American area in per capita production of services was even more marked. At the end of the period under review North America exceeded the average West European level by 160 per cent and was more than three times the level of the other capitalist countries. In this sphere, as in the other main sectors of the GDP, a tendency is discernible in turn toward a levelling up, which is clearly displayed in the consolidation of its importance in the economy of modern capitalism, per capita regional and national indicators included. But even allowing for a very probable gradual intensification of this trend, the existing considerable gap in levels of development will remain a feature of the economies of capitalist countries for the last twenty years of the century.

A certain smoothing out of previously formed substantial differentiation of the role and place of trade and commerce in extended reproduction has also been characteristic of the postwar development of these economies. It has come about, as a rule, through a decline in the proportion of trade in their GDP, where it was traditionally high, and through a rise where it was comparatively low. In spite of the gradual convergence of the per capita indicators in this sector, however, more than half of the added value created in the trade and commerce of the developed capitalist countries taken together came at the end of 70s from the North American region, and the per capita gap, moreover, remained particularly wide. Analysis of the postwar trends allows us to conclude that the weight of trade and commerce may hardly change essentially on an average. For most countries, apparently, it will remain between one-sixth and one-seventh of the GDP.

The regional dynamics of transport and communications in the industrial centres had much in common with that of the growth of trade and services. Given the increasing unevennessof growth in the various countries and in these sectors under the scientific and technical revolution, a levelling tendency forced its way through, which led to a gradual closing of the gap between them in both absolute terms and per capita indices of production. At the end of the period reviewed, however, nearly as much was produced in value terms in the North American region as in all the developed countries of the other regions.

The tendencies to 'self-movement' of the main sectors of production in world capitalism's centres considered in this section bring out, in their aggregate, several very important shifts in its postwar structure. In essence they are various forms of the manifestation of broader processes reflecting both the course of the unevenness of capitalism's economic development and the objective needs of the further growth and intcrnationalisation of society's productive forces. Concrete analysis of these trends undoubtedly remains topical for a description of the underlying processes taking place within this mode of production in our time.

These needs have largely determined the regional and sectoral shifts in the postwar production of the developing countries in the ever deepening crisis of world capitalism's former colonial relations. In the stage of imperialism, as we know, the sphere of operation of the law of uneven economic and political development hegan to embrace not only its centres hut also their agrarian appendages drawn by the colonial powers and foreign capital into the maelstrom of international economic relations. In October 1917 Lenin insistently stressed: As if all peoples were *equally* drawn into this one world economy! As ii' there existed no relationship of *bondage* between llie uncivilised and the 'civilised' peoples precisely *on the basis* of 'all peoples' being drawn 'into one world economy'.

This postulate of his has acquired special significance and topicality with the break-up of imperialism's colonial system. But a wide range of matters arises here, stemming from the specilic features of the manifestation of the law of uneven development in the former colonial periphery, matters that call for special, comprehensive analysis.

The Agrarian and Raw Material Producing Countries

The winning of political independence by the bulk of the countries of the colonial world in the postwar period opened up fundamentally new opportunities for them to fight to step up their national development, and for socio-economic progress and equality in international relations against imperialist diktat and exploitation by expatriate monopoly capital. As Leonid Brezhnev said at the 25th Congress of the CPSU:

Glancing at the picture of the modern world one cannot help noticing the important fact that the influence of states that had only recently been colonies or semi-colonies has grown consider ably.

It may definitely be said about the majority of them that they are defending their political and economic i ights in a struggle against imperialism with mounting energy, striving to consolidate their independence and to raise the social, economic and cultural level of their peoples.

In these conditions important changes are occurring or coming into life in the structure and distribution of the developing world's social production, which are putting their stamp on the processes that govern the peculiarities of the former colonies' and semi-colonies' position in the present international division of labour, including their mounting struggle to lay the foundations of their economies. In the 50s through the 70s the physical volume of their industrial production as a whole rose by more than 450 per cent, so that now, in contrast, to the earlier stages of their economic history, industrial goods, especially manufactures, are beginning to predominate in the material production of most of them. At the end of our period somewhat more was produced by manufacturing industry in value terms (in 1970 prices) than in agriculture, and roughly four times as much as in mining.

As the general economic capability, and especially the industrial potential, of these countries grew there was an inevitable consolidation of the demarcation of class forces, a deepening of the heterogeneity of their social development, and a growth of class struggle. For the immense majority, at the same time, that were to one degree or another within the orbit of the laws of the world capitalist economy, it remains characteristic that the tasks of the anti-feudal revolution have not been completed and the socio-economic survivals of the colonial period are very burdensome. Many of the new sovereign states had begun to carry through radical reforms during the liberation struggle that were both antifeudal and anti-capitalist. In thatsituation the differentiation of their paths and levels of economic development greatly broadened.

In spite of essential differences of that kind, however, all the emancipated countries are part of the agrarian and raw material periphery of the industrial centres of capitalism. Even the most economically advanced of them are still mainly exporters of raw materials arid importers of machinery, equipment, and other industrial goods. They 158 continue to he in an unequal position within the capitalist world system and subjected to exploitation by expatriate monopoly capital, which is why they are being urgently faced with a number of common internal and external economic problems.

The most important trends in the changing postwar structure of the periphery's social production have been analysed above. How, however, was this production distributed among the developing countries in the first postwar decades?

As will be clear from the graphs the growth rates of aggregate production were very fast in Latin America, where most of the countries had achieved national independence and taken the road of independent capitalist development long before break-up of the colonial system began. As a result the postwar years have been marked by a gradual raising of the role of Latin America in the aggregate product of the developing world, from 40 to 44.5 per cent, so that it is now greater than that of the Asian region although the latter had nearly four times its population at the end of the period surveyed.

This growth did not come about through the contribution of all Latin American countries, but occurred on a background of a marked imbalance of their economic development and was mainly determined by a few leading countries, primarily Brazil and Mexico, for which much higher growth rates were typical. At the end of the 70s more than 60 per cent of the aggregate product of the region was produced in these two countries alone, whereas they produced around 40 per cent of it at the beginning of the 50s. At the same time certain countries (Haiti, Honduras, Barbados, etc.), still caught up in heavy chains of dependence on expatriate commodity monopolies, remained among the most economically 159 backward appendages of the world capitalist economy, with the usual low growth rates typical of semi-colonial countries.

A steady deepening of the unevenness of Hie intor-regional. and especially of the intraregional, distribution of production has become typical of the various groups of commodityproducing countries in the other continents. In Africa, where colonialism succeeded in retaining key positions in many areas for a number of years in the postwar period, the course of economic development has proved to be relatively slow, so that the weight of the continent in the aggregate social product of developing countries has had a clear tendency to fall, despite the fact that the total GDP of African countries has risen substantially.

As national independence has been consolidated and African countries have thrown off the yoke of colonial exploitation, however, their growth rates, it must be noted, as in other areas of the developing world, have gradually risen. They have been highest, as a rule, in the Arab countries in North Africa, where the positive consequences of the breakup of the colonial system made themselves felt much earlier than in Tropical Africa. The rapid increase in income from exploiting their oil wealth, moreover, provided some of them with favourable material conditions for their struggle for quicker economic development.

There has been an ever wider difference in levels and rates of development in Asian countries, where around two-thirds of the developing'world's population lives. Most of Ihem, which were in forced dependence on a handful of European powers during nearly the whole history of the forming of the world capitalist economy, were able to win national independence at the start of the postwar period. In the 50s they had already begun an active fight for accelerated growth of their previously stagnating productive forces. The mean annual growth rates of the GDP of all the developing countries of Asia reached 4.5 per cent in that decade, 5.1 per cent in the 60s, and around 6 per cent in the 70s. This accelerated growth was due to no small degree in recent times, however, to the 'oil boom' in the

Near and Middle East. The real volume of the GDP (in value terms) increased more than sixfold in that region in the 50s through the 70s, and was more than one-third of the GDP of all the developing countries of Asia, although its population at the end of the period was only 7 per cent of the continent's total. At the same time the GDP of all the other developing countries in South and South-East Asia, which have immense human resources, rose less than threefold.

If we examine these processes through the prism of the data on population growth, the generalisations made above, stemming from a comparative description of the regional dynamics of production, obviously need to be made more precise, and corrected. The successive changes in the indicators of per capita GDP for the separate regions and countries enable the scale of the shifts in the geographical distribution of their productive forces to be brought out more clearly.

In almost all the groups of countries in the table, there is a quite marked tendency, albeit extremely varied in dynamics and significance, toward a differentiation of population growth rates, in addition to a sharpening of the differences in dynamics of production. This trend characterised one of the inevitable consequences of the break-up of the colonial system, and could not help having an essential influence on the ratio of economic potentials, and consequently on the mutually interlocked processes of economic development.

In the Asian region a particularly wide gap has been formed in the dynamics of per capita production of the aggregate social product between the oil-producing countries of the Near and Middle East and the overwhelming majority of the countries of South and South-East Asia.

Similar processes have been occurring in Africa, for all its special features. The growing uncvenness of development has been deepened there, even more than in Asia, by considerable differences in rates of population growth. In some cases these differences, it is true, have encouraged a convergence of the dynamics of per capita production in the separate groups of countries. That is confirmed, in particular, by comparison of the lines of postwar development of North and Tropical Africa. Population has grown faster in the former, which has led to a certain levelling out of the aggregate indicators of per capita growth of production used in the table. At the same time there have been quite different trends within each of these regions.

In the biggest North African country, Egypt, for example, there was an obvious slowing of the growth rates both of production and of population in the 60s and 70s. The annual rates of per capita growth of GDP in those years were 1.5 per cent in Egypt and only 0.2 per cent in Sudan. In other words, they were lower than in many of the countries of Tropical Africa whose economies on the whole still suffer maximally from the burden of harmful survivals of the colonial period compared with other developing countries. Although some of the new states south of the Saharajiave achieved a relatively high growth of production (Kenya, Gabon, Ivory Coast, Nigeria, Malawi, etc.), most of them still have extremely slow and very unstable rates of development. In quite a few countries, moreover, the main indices of the dynamics of GDP growth have been lower in recent decades than the growth of population (Burundi, Chad, the Central African Empire, Uganda, Upper Volta).

The long-term trends in the distribution of production in the former colonial regions are primarily governed by the resultant tendencies of the development of a few of the biggest countries (as regards area, population, and natural 164
resources). The ratio of their economic capabilities is therefore important in a generalised description of intraregional processes, including the contradictory consequences of the 'demographic explosion' in the Third World.

In Latin America, for example, GO per cent of the inhabitants and nearly two-thirds of the aggregate GDP of the whole region were concentrated in Brazil, Mexico, and Argentina. Among the 'Big Three' production developed at the lowest rates in Argentina in the postwar years, but the comparatively slow dynamics of its population growth led to its per capita production, for example, exceeding Brazil's in the 60s and Mexico's in the early 70s. It is typical that the economies of several small countries (Uruguay, El Salvador, Paraguay, Haiti, Honduras, and others) wore below the average of the developing countries of South and SouthEast Asia as regards the same indicators in those years, not to mention of bulk of the emancipated countries of Africa. The examples given, whose number could be greatly extended, very clearly confirm that there has been a steady sharpening of the unevenness of economic development of the former colonies and semi-colonies in recent decades. This irresistibly intensifying process, however, in no way refutes the need (in addition to the study of the features of the economic growth of each country) for a comprehensive systems approach to analysis of the summary trends of their interconnected development within the whole developing world. Such an approach helps, on the one hand, to bring out the objectively operating trends in which the extremely contradictory and isolated processes and phenomena characterising the shifts in the distribution of the productive forces of the separate countries and their regional groups are ultimately reduced, and on the other hand, gives the possibility of better understanding and appreciating the specific features of their development.

In this connection comparative analysis of the long-term changes in the industrial structure of the social production of countries in Asia, Africa, and Latin America is becoming of paramount importance.

As these countries are industrialised this process will undoubtedly hec07iie more and more pronounced. Its effect gives grounds for supposing that even before the end of this century the share of agriculture in the GDP of most of the developing countries of Asia and Africa may fall by 50 to 50 per cent and reach the level attained in Latin America al the beginning of the 80s. At the same time the proportion of agriculture in the GDP of the Latin American region as a whole will also probably diminish, gradually approaching the present average level of the industrial capitalist countries (around 4 per cent). An inevitable consequence of that will be the pushing of increasing numbers of the rural population into the cities.

Another very important trend determining botli the current structure of changes in the production of the developing countries and their probable future structure has become the general though not uniform rise-in the role of industry in their GPP. The scale of its growth, moreover, has been quite regularly higher in the economically more backward countries. Thus the proportion of industrial output in the aggregate product for the period considered in the graphs increased in the African region by 120 per cent (from 12 to 26 per cent), in the Asian region by almost 100 per cent (from 14 to 27 per cent), and in the Latin American region by almost a third (from 24 to 31 per cent).

As a result of the increasing imbalance of this growth, a trend toward a certain convergence of the significance and place of industry in the general structure of these regions' production has been noted in the postwar period. The proportion of industry in the GDP of countries in South and South-Easl Asia and in Africa was 50 per cent less than in Latin America at the beginning of the 50s; at the beginning of the 80s the gap had noticeably closed and was estimated now at 1:1.3 for the former and 1:1.2 for tho latter.

This trend will intensify, in all probability, in the future, and characterise one of the tendencies of developing countries' uneven economic growth. It can be expected that the weight of industry in the GDP of most Asian and African] countries will come close to tho present level in the leading Latin American countries in the next two or three decades, while the latter will reach the average level of the present-day industrial centres of capitalism.

These very general estimates provide the dnta needed, but only the initial data, for studying the significance and place of industry in the regional economies. They naturally smooth over tho vast differences in structure and sectoral dynamics of the industrial production of the separate regions and countries. A more and more concrete analysis of the separate industries is a central task of the part that follows, but here we must bear in mind that these processes of industrial development in agrarian and raw material producing countries are largely linked with rapid growth of mining, whose output is mainly sold on the world capitalist market, and in fact is still really controlled to one degree or another by expatriate monopoly capital. The rapid growth of the mining of minerals, and in particular of fuel, has played no small role in the regions being considered in the postwar rise in the proportion of industry in the gross product.

It would not be legitimate, however, to conclude from this that the steady raising of the role of industry in the gross product of developing countries is more or less wholly 168 linked with the working of minerals. The rapid development of their manufacturing industries in recent decades has had an ever increasing effect, especially the development of heavy industry, the volume of whose production rose on the whole by almost 150 per cent in 1968–80. At the beginning of the 80s the heavy industries of all developing countries were producing 30 per cent more output in value terms (in added value in 1975 prices) than the light industries. There is little doubt that their basic industries will also grow al faster rates in the next few decades, which will further a gradual raising of their role in industrial production.

For a comparative estimate of the long-term trends bringing out both the general and the particular in the production sphere of the main developing regions of capitalism, it is essential to compare the mean annual growth rates of their main industries (including per capita rates) over some considerable period. Because of the inadequacies of the economic statistics of developing countries it is very difficult to bring out these growth rates in summary form (in contrast to developed countries). The comparison can only be attempted on the basis of very approximate calculations, which allow us to characterise the most general trends of the sectoral shifts with a certain degree of accuracy. At the same time they can be used as a starling point of sorts for studying the specific features of the development of production in the separate countries and groupings in each of the regions.

In agriculture, which is still in a slate of deep, protracted depression in the developing world as a whole, Latin America and most of the countries of the Near and Middle East have displayed relatively faster growth, so that a line toward an increase in their share in developing countries' total production of farm produce has begun to show., with a corresponding drop in the role of South and South-East Asia and Africa. As a result they have become further differentiated as regards per capita farm production. By the beginning of the 80s the Latin American countries on the whole considerably surpassed the countries of Asia and Africa in this respect. It is unlikely that there will be any closing of this gap in the near future. The fight to eliminate the extreme backwardness of agriculture will undoubtedly remain one of the most pressing socio-economic problems of most of the countries of the former colonial world in the coming decades. It is the slow development of their agriculture, compared with population growth, that is the decisive cause of the consistent tendency manifested since the war toward a widening of the gap in GDP per capita between the developing and developed countries of the capitalist world.

In the 60s and 70s, however, this trend began to be more and more actively countered by processes developing in most of the other sectors of the economies of the former colonial periphery of capitalism, and primarily in industry. The international statistics clearly indicate that the per capita growth rates of industrial production in economically most backward regions of capitalism's periphery have noticeably exceeded the industrial centres in per capita growth rales of industrial production. In the developing countries of Asia per capita industrial production increased by a factor of 4.3 from 1950 to 1978, in Africa by 3.5, and in the developed capitalist countries taken together by 2.6. The countries of Latin America also showed a rather faster growth in this respect, on an average, than the main economic regions of capitalism.

In present conditions there has begun lo be a tendency toward a certain closing of the gigantic, gap in per capita industrial production formed under colonialism, together with 171 a gradual increase in the weight of the agrarian and raw material producing countries in the industrial production of the non-socialist world. In Hie very early 50s this gap was estimated at 1:35 for Asian countries, 1:32 for Africa, and 1:5.5 for Latin America. By the end of the 70s it had been clearly reduced and was expressed by the following approximate ratios: Asia 1:24, Africa 1:27, and Latin America 1:5.3. In the middle of 70s the substantial slowing of industrial growth rates in the developed countries connected with the greatest postwar cyclic crisis of the capitalist world economy fostered a consolidation of this trend.

The general course of this process leads to th,e conclusion that it is acquiring a quite stable, irreversible character in Ihc present situation. Being a natural consequence of the break-up of the colonial system, it has already begun to have (and in all probability will continue to have) a growing impact on the steady deepening of the crisis of the unequal system of international capitalist division of labour built up on a colonial basis.

At the same time the facts adduced confirm the extreme complexity and difficulty of the problem of overcoming the age-old industrial backwardness of the former colonies and semi-colonies, especially with their constant exploitation by expalriate monopoly capital. The solution of such a major and objectively inevitable hislorical task will obviously require several decades and will continue for a long time to be one of the urgent problems ol the light of the peoples of the liberated countries for final liquidation of the heavy legacy of the colonial period and imperialist domination of their economies.

To reduce their backwardness by even two-thirds from the present mean per capita industrial production of the developed countries, for instance, it would take the countries of South and South-East Asia around a quarter of a century at the rates of real industrial growth of the 60s and 70s, Africa more than a quarter of a century, and Latin America around 30 years. Then, however, per capita output of industrial goods in each of these regions would the respectively one-eighth, a ninth, and about a third below the level already reached by Ihe advanced capitalist countries at the end of the 70s. If that level remains unchanged in the future, then in that hypothetical case the way out for most developing countries would have lo be a 15-fold or 16-fold expansion of the volume of industrial production within the present-day world capitalist economy.

These figures cannot, of course, serve as a generalised criterion of a quantitative, let alone a qualitative, estimate of the future parameters and goals of the industrial development of all the developing countries of Asia, Africa, and Latin America. In fact, however, they help determine only the scale of the industrial backwardness of developing countries. It inevitably follows from them that the struggle for a cardinal solution of the problem in the present transitional epoch will give rise to an urgent need to develop a long-term strategy and-strictly scientific approach to the global outlook for industrialisation of the former colonies and semi–colonies on quite other principles of social and world economic relations compared with those historically built up by capitalism.

The postwar processes in the agrarian and industrial spheres have consequently had a different effect on the resultant ratio of per capita GDP in the groups of countries under consideration. In the agrarian sphere they have fostered a widening of the economic gap, and in the industrial sphere a narrowing. But since the role of the agrarian sector in the GDP of developing countries is more significant than in the centres of capitalism, and the industrial sector less significant, the gap between them in per capita output of the main sectors (agriculture, industry, and building) has also continued to widen in recent decades, and for Asian and African countries was more than 10:1 at the beginning of the 80s against 8:1 in the first postwar years. In other words, at the turn of the decade to the 80s, per capita production in value terms (in comparable prices) was much lower in developing countries compared with capitalist ones, than at the turn of the 40s and 50s.

As for industry, its growth in developing countries has still not led to any real qualitative shifts in the liquidation of age-old industrial backwardness. The struggle to lay an industrial foundation for their national economies is still more or less, as a rule, in the initial stage. In spite of a narrowing of the gap in per capita industrial output, they now produce dozens of times less product than the average for capitalist countries. The most powerful, technically advanced branches of production, moreover, that can more efficiently exploit the fruits of the present scientific and technical revolution are concentrated in the latter. That is affecting the dynamics of productivity in the industry of both groups of countries. In the 60s and 70s its annual growth rates in the economic centres of capitalism were double that in the developing countries.

There has been a quite stable tendency toward a gradual, tliough slow, rise in the proportion of circulation and services in the GDP of the developing world, which rose perceptibly in 1900–80. As a result of the rapid expansion of this sphere its weight in the GDP of the Asian and African regions rose by a quarter and nearly a third respectively in the 50s through the 70s. It rose a little as well in Latin America on the whole and constituted more than half of that region's GDP at the end of the 70s. In other words, there was a certain convergence of structural proportions in the sectors of non-materialproduction of the main developing regions, but the historical underdevelopment of these sectors still remains a distinguishing sign of the economic backwardness of the vast majority of the former colonies and dependent countries of these regions.

The proportion of developing countries in the aggregate production of the services and circulation sphere of the world capitalist economy is beginning lo grow. At the start of llio 80s it was 10 per cent against 11 per cent in the early 50s. The per capita gap between them and the developed countries, it is true, not only did not in fact diminish but even widened a bit on the whole, and was estimated at 15.5:1 in the late 70s, i.e. was considerably wider than in the sectors of material production.

For the main sectors of the circulation and services sphere this gap has not built up in the same way; in practice it has always been narrower in trade. But it is in that sector, as in agriculture, that the capitalist countries continued steadily to outpace the developing countries in per capita indices of growth, so that, although the proportion of the latter in the total value of the home trade of the capitalist world lias risen since the war (from one-seventh to one-sixth), there has been another trend in per capita terms While the value of per capita trade (in constant 1975 prices) in the GDP of capitalist countries was 12 times as much as in developing countries at the beginning of the 60s, it was 16 times as much at the start of the 70s. This trend above all determined the effect in the circulation and services sphere of the tendency toward a further widening of the general economic gap in per capita GDP between the two groups of countries.

The intensification of this trend also to no small extent was promoted by processes taking place in the sectors producing so-called services. Developing countries' backwardness in this respect has long been deeper than in the other sectors of the sphere of social production we are considering. At the turn of the 50s and 60s it corresponded approximately to the level of their industrial backwardness. At that time less than one-tenth of the value of the services created in all capitalist countries was produced in developing countries, and the gap per capita was nearly 20:1. In recent years there have been changes in various directions in these ratios. The rapidly increasing dynamics of the growth of services in developing countries has led to their share rising in absolute terms to one-seventh at the end of the 70s. But in per capita terms the gap continued to widen for a long time, and in the early 70s could be expressed as 19.5:1, which was mainly due to the backwardness in this 175 area of the economically less developed and more densely populated countries of the African and Asian regions.

In the second half of the 70s alone, especially during the world cyclic crisis of 1974–75, a long-term consequence of which was a marked slowing of the growth rales of services in the centres of capitalism, did this gap begin to narrow a bit; at the beginning of the 80s it was expressed by a ratio of 17.5:1 (in 1975 prices). In other words, it has remained wider than in industry. The extreme backwardness of the services sphere governs several of the paramount aims of the long-term socio-economic strategy being developed by developing countries today in their striving to attain the maximum possible speeding up of the rates of social progress in present-day conditions and to raise the standard of living of their populations.

The only major group of services and circulation sectors in which developing countries managed to get a relatively stable increase in the dynamics of per capita production over that of the developed countries in the postwar decades was transport and communications; the grand total of per capita production in them increased by a factor of 3.4 against 2.4 in the developed countries.

There was a corresponding ri se (from 11 to 16 per cent) in developing countries' weight in the aggregate value of the product of transport and communications in the world capitalist economy. In per capita terms their mean annual rates of development have been higher of late than those of the industrial regions, so that in the area of the infrastructure, and in industry, there has begun to be a relatively new and apparently promising trend toward a certain narrowing of the gap in per capita indices, though it still remains extremely wide. If the present dynamics of the growth of production and population is maintained the gap should be halved at least on an average in the coming two or three decades in each of the main regions of the developing world, but its ultimate closing will undoubtedly still remain an urgent task for a long time.

At the same time there was a clearly increasing unevenness in the development of transport and communications in the developing countries, which, as the figures given above show, 176 was organically characteristic of all the other sectors of tho circulation and.services sphere. Its dynamics will also undoubtedly be a close function of the degree and scale of development of material product ion and of the population grovyth rates in the various regions and countries of the developing world. On the whole, however, one can say with confidence that as the productive forces rise the weight of this sphere in the social production of these countries will gradually increase although it will still lag greatly behind the corresponding average indicator for the industrialised countries for several decades.

The wide gap between the developing countries and the centres of capitalism in all the decisive areas of economic activity consequently remains one of the objective realities of our time. Formed long ago, during the building of the world capitalist economy on a colonial basis, it now characterises many of the most important features of that economy, and of tho whole system of its postwar international relations.

The developing countries of Asia, Africa, and Latin America, given the existing differences in their levels of sociopolitical and economic development, arc inevitably coming up against a need to deal with several common problems, vitally important for all developing countries, on both an international and a national scale. They include the very intricate set of each country's national problems, but ultimately tho struggle to cope witli all of them, taken together is objectively directed to eliminating the pernicious surviv' als of the colonial period from the affairs of human society" and to emancipating the developing countries from neoco, lonialism and continuing capitalist exploitation.

At the same time, as will be clear from the facts adduced above, the gap in levels of economic developm ent remains extremely wide, in spite of considerable progress by the developing countries in coping with these problems, and moreover retains a tendency to widen in several determinant indicators, above all in gross product per capita. The relations of inequality and rapacious exploitation of the natural and human resources of the developing countries by expatriate monopoly capital prevailing in the world capitalist economy are furthering maintenance of this trend, even since the break-up of imperialism's colonial system. The capitalist social system itself not only bears responsibility for the socio-economic backwardness of the peoples of the former 177 colonial world, but the exploiter system of international division of labour created by it largely continues today to fetter their productive forces. The basic patterns of the world capitalist economy still operate in the direction detrimental to the developing countries.

The striving of the developing countries to spread the liquidating of colonialism to the economic sphere, to put an end to their exploitation by the industrial powers of the West, and to achieve creation of the necessary external and internal conditions for attaining a level of development in the foreseeable future corresponding to the needs of today, is quite justified. At the same time all the nations of the earth without exception have an interest in the speediest attainment of these aims. The struggle for a steady acceleration of socio-economic and cultural progress in the former colonies and semi-colonies is therefore becoming an inseparable part of the fight of all progressive mankind for detente and the consolidating of lasting peace, for further development of the scientific and technical revolution for peaceful purposes, and for a favourable solution of the other global problems of today on which the whole subsequent course of world development to an enormous extent depends.

In his day Karl Marx concluded that mankind thus inevitably sets itself only such tasks as it is able to solve, since closer examination will always show that the problem itself arises only when the material conditions for its solution are already present or at least in the course of formation.

In the present historical situation more favourable objective conditions are being created than ever before for the countries of Asia, Africa, and Latin America that have liberated themselves from the imperialist yoke, or are in the course of doing so, for a successful struggle to overcome the backwardness inherited by them from the past in all sectors of 178 the economy and oilier areas of social affairs. But every kind of obstacle is still being put in llio way of this bard, long, stubborn light with the world capitalist economy by the former colonialists and expatriate monopoly capital. In these conditions the developing countries are uniting their efforts more and more firmly in the anti-imperialist movement to establish a new international economic order.

The Soviet Union and other socialist countries are supporting these efforts in every way possible and demonstrating their solidarity with the progressive aims of the struggle of the peoples of the developing world for a better future. As the Central Committee's Report to the 26'th Congress of the CPSU stressed: the CPSU will consistently continue the policy of promoting cooperation between the USSR and the ncwly-freo countries, and consolidating the alliance of world socialism and the national liberation movement.

The long-term trends reviewed in this chapter thus not only help us better to understand some of the general results of the changes that have already taken place in the structure of the distribution of the productive forces of the two groups of countries of the postwar capitalist world, but also to distinguish sqveral objective premises for analysing the probable outlook for subsequent structural shifts leading to aggravation of the internal contradictions and crisis phenomena in the capitalist economy. Study of these trends obviously calls for a need to allow in every way for the effect on them of the spontaneously operating laws of the cyclic development of the capitalist mode of production.

9

Ecology and the Transition from Capitalism to Socialism

The transition from capitalism to socialism is the most difficult problem of socialist theory and practice. To add to this the question of ecology might therefore be seen as unnecessarily complicating an already intractable issue. I shall argue here, however, that the human relation to nature lies at the heart of the transition to socialism. An ecological perspective is pivotal to our understanding of capitalism's limits, the failures of the early socialist experiments, and the overall struggle for egalitarian and sustainable human development.

My argument has three parts. First, it is crucial to understand the intimate connection between classical Marxism and ecological analysis. Far from being an anomaly for socialism, as we are often led to believe, ecology was an essential component of the socialist project from its inception—notwithstanding the numerous later shortcomings of Soviet-type societies in this respect. Second, the global ecological crisis that now confronts us is deeply rooted in the "world-alienating" logic of capital accumulation, traceable to the historical origins of capitalism as a system. Third, the transition from capitalism to socialism is a struggle for sustainable human development in which societies on the periphery of the capitalist world system have been leading the way.

CLASSICAL MARXISM AND ECOLOGY

Research carried out over the last two decades has demonstrated that there was a powerful ecological perspective in classical Marxism. Just as a transformation of the human relation to the earth was, in Marx's view, an essential presupposition for the transition from feudalism to capitalism, so the rational regulation of the metabolic relation to nature was understood as an essential presupposition for the transition from capitalism to socialism. Marx and Engels wrote extensively about ecological problems arising from capitalism and class society in general, and the need to transcend these under socialism. This included discussions of the nineteenth-century soil crisis, which led Marx to develop his theory of metabolic rift between nature and society. Basing his analysis on the work of the German chemist Justus von Liebig, he pointed to the fact that soil nutrients (nitrogen, phosphorus, and potassium) were removed from the soil and shipped hundreds and thousands of miles to the cities where they ended up polluting the water and the air and contributing to the poor health of the workers. This break in the necessary metabolic cycle between nature and society demanded for Marx nothing less than the "restoration" of ecological sustainability for the sake of "successive generations."

In line with this, Marx and Engels raised the main ecological problems of human society: the division of town and country, soil depletion, industrial pollution, urban maldevelopment, the decline in health and crippling of workers, bad nutrition, toxicity, enclosures, rural poverty and isolation, deforestation, human-generated floods, desertification, water shortages, regional climate change, the exhaustion of natural resources (including coal), conservation of energy, entropy, the need to recycle the waste products of industry, the interconnection between species and their environments, historically conditioned problems of overpopulation, the causes of famine, and the issue of the rational employment of science and technology.

This ecological understanding arose from a deep materialist conception of nature that was an essential part of Marx's underlying vision. "Man," he wrote, "*lives* from nature, i.e. nature is his *body*, and he must maintain a continuing dialogue with it if he is not to die. To say that man's physical and mental life is linked to nature simply means that nature is linked to itself, for man is a part of nature." Not only did Marx declare in direct opposition to capitalism that no individual owned the earth, he also argued that no nation or people owned the earth; that it belonged to successive generations and should be cared for in accordance with the principle of good household management.

Other early Marxists followed suit, although not always consistently, in incorporating ecological concerns into their analyses and embodying a general materialist and dialectical conception of nature. William Morris, August Bebel, Karl Kautsky, Rosa Luxemburg, and Nikolai Bukharin all drew on ecological insights from Marx. The Ukrainian socialist Sergei Podolinsky's early attempt at developing an ecological economics was inspired to a considerable extent by the work of Marx and Engels. Lenin stressed the importance of recycling soil nutrients and supported both conservation and pioneering experiments in community ecology (the study of the interaction of populations within a specific natural environment). This led to the development in the Soviet Union in the 1920s and early 1930s of probably the most advanced conception of ecological energetics or trophic dynamics (the basis of modern ecosystem analysis) in the world at the time. The same revolutionary-scientific climate produced V. I. Vernadsky's theory of the biosphere, A. I. Oparin's theory of the origin of life, and N. I. Vavilov's discovery of the world centers of germplasm (the genetic sources of the world's crop plants). In the West, and in Britain in particular, leading scientists influenced by Marxism in the 1930s, such as J. B. S. Haldane, J. D. Bernal, Hyman Levy, Lancelot Hogben, and Joseph Needham, pioneered in exploring the dialectics of nature. It is even possible to argue that ecological science had its genesis almost entirely in the work of thinkers on the left (socialist, social democratic, and anarchist).

Obviously not all major figures or all developments in the socialist tradition can be seen as ecological. Soviet Marxism succumbed to an extreme version of the productivism that characterized early twentieth-century modernity in general, leading to its own version of ecocide. With the rise of the Stalinist system the pioneering ecological developments in the Soviet Union were largely crushed (and some of the early ecologically oriented Marxists such as Bukharin and Vavilov were killed). Simultaneously, a deep antipathy to natural science emerging out of an extreme negation of positivism led to the abandonment of attempts to theorize the dialectics of nature in Western Marxism, seriously weakening its link to ecology—

though the question of the domination of nature was raised by the Frankfurt School as part of its critique of science. If today socialism and ecology are once again understood as dialectically interconnected, it is due both to the evolution of the ecological contradictions of capitalism and the development of socialism's own self-critique.

CAPITALISM'S WORLD ALIENATION

The key to understanding capitalism's relation to the environment is to examine its historical beginnings, i.e., the transition from feudalism to capitalism. This transition was enormously complex, occurring over centuries, and obviously cannot be fully addressed here. I shall focus on just a few factors. The bourgeoisie arose within the interstices of the feudal economy. As its name suggests, the bourgeoisie had its point of origin as a class primarily in the urban centers and mercantile trade. What was necessary, however, in order for bourgeois society to emerge fully *as a system*, was the revolutionary transformation of the feudal mode of production and its replacement by capitalist relations of production. Since feudalism was predominantly an agrarian system, this meant of course transformation of agrarian relations, i.e., the relation of workers to the land as a means of production.

Capitalism therefore required for its development a new relation to nature, one which severed the direct connection of labor to the means of production, i.e., the earth, along with the dissolution of all customary rights in relation to the commons. The *locus classicus* of the industrial revolution was Britain, where the removal of the workers from the land by means of expropriation took the form of the enclosure movement from the fifteenth to the eighteenth centuries. Under colonialism and imperialism an even more brutal transformation occurred on the outskirts or the external areas of the capitalist world economy. There all preexisting human productive relations to nature were torn asunder in what Marx called the "extirpation, enslavement and entombment in mines of the indigenous population"—the most violent expropriation in all of human history.

The result was proletarianization within the center of the system as masses of workers were thrown out of work and moved to the city. There they were met by the capital being amassed through organized robbery, giving rise to what Marx called "modern industry." Simultaneously, various forms of servitude and what we now call precarious work were imposed on the periphery, where social reproduction was always secondary to the most rapacious imperialist exploitation. The surplus forcibly extracted from the periphery fed industrialization at the center of the world economy.

What made this new system work was the incessant accumulation of capital in one cycle after another, with each new phase of accumulation taking the last as its starting point. This meant ever more divided, more alienated human beings, together with a more globally destructive metabolism between humanity and nature. As Joseph Needham observed, the "conquest of Nature" under capitalism turned into "the conquest of man"; the "technological instruments utilized in the dominance of Nature" produced "a qualitative transformation in the mechanisms of social domination."

There is no doubt that this dialectic of domination and destruction is now spiraling out of control on a planetary scale. Economically, overall inequality between the center and periphery nations of the world system is increasing together with the intensification of class inequality within each capitalist state. Ecologically, the world's climate and the life-support systems of the entire earth are being transformed by a process of runaway global warming.

In addressing this planetary environmental problem it is useful to turn to Hannah Arendt's concept of "world alienation," introduced fifty years ago in *The Human Condition*. "World alienation" for Arendt began with the "alienation from the earth" at the time of Columbus, Galileo, and Luther. Galileo trained his telescope on the heavens, thereby converting human beings into creatures of the cosmos, no longer simply earthly beings. Science seized on cosmic principles in order to obtain the "Archimedean point" with which to move the world, but at the cost of immeasurable world alienation. Human beings no longer apprehended the world immediately through the direct evidence of their five senses. The original unity of the human relation to the world exemplified by the Greek polis was lost.

Arendt noted that Marx was acutely aware of this world alienation from his earliest writings, pointing out that the world was "denatured" as all natural objects—the wood of the wood-user and the wood-seller—were converted into private property and the universal commodity form. Original or primitive accumulation, the alienation of human beings from the land, as Marx described it, became a crucial manifestation of world alienation. However, Marx, in Arendt's view, chose to stress human self-alienation rooted in labor rather than world alienation. In contrast, "world alienation, and not [primarily] self-alienation as Marx thought," she concluded, "has been the hallmark of the modern age."

"The process of wealth accumulation, as we know it," Arendt went on to observe, depended on expanding world alienation. It "is possible only if the world and the very worldliness of man are sacrificed." This process of the accumulation of wealth in the modern age "enormously increased human power of destruction" so "that we are able to destroy all organic life on earth and shall probably be able one day to destroy even the earth itself." Indeed, "Under modern conditions," she explained, "not destruction but conservation spells ruin because the very durability of conserved objects is the greatest impediment to the turnover process, whose constant gain in speed is the only constancy left wherever it has taken hold."

Arendt had no final answers to the dire problem she raised. Despite tying world alienation to a system of destruction rooted in wealth accumulation, she identified it with the development of science, technology, and modernity rather than capitalism as such. World alienation in her view was the triumph of *homo faber* and *animal laborans*. In this tragic conception, her readers were called upon to look back to the lost unity of the Greek polis, rather than, as in Marx, toward a new society based on the restoration at a higher level of the human metabolism with nature. In the end world alienation for Arendt was a Greek tragedy raised to the level of the planet.

There is no doubt that the concrete manifestations of this world alienation are evident everywhere today. The latest scientific data indicate that global emissions of carbon dioxide from fossil fuels experienced a "sharp acceleration...in the early 2000s" with the growth rate reaching levels "greater than for the most fossil-fuel intensive of the Intergovernmental Panel on Climate Change emissions scenarios developed in the late 1990s." Further, "the mean global atmospheric CO2 concentration" has been increasing "at a progressively faster rate each decade." The most rapid acceleration in emissions has been in a handful of emergent industrializing countries such as China, but "no region" in the world is currently "decarbonizing its energy supply." All ecosystems on earth are in decline, water shortages are on the rise, and energy resources are becoming more than ever the subject of global monopolies enforced by war.

The "man-made fingerprint of global warming" has been detected "on 10 different aspects of Earth's environment: surface temperatures, humidity, water vapor over the oceans, barometric pressure, total precipitation, wildfires, change in species of plants and animals, water run-off, temperatures in the upper atmosphere, and heat content in the world's oceans." The cost now descending on the world if it doesn't radically change course is a *regression* of civilization and life itself beyond comprehension: an economy and ecology of destruction that will finally reach its limits.

SOCIALISM AND SUSTAINABLE HUMAN DEVELOPMENT

How are we to meet this challenge, arguably the greatest that human civilization has ever faced? A genuine answer to the ecological question, transcending Arendt's tragic understanding of world alienation, requires a revolutionary conception of sustainable human development—one that addresses both human self-estrangement (the alienation of labor) and world alienation (the alienation of nature). It was Ernesto "Che" Guevara who most famously argued in his "Man and Socialism in Cuba" that the crucial issue in the building of socialism was not economic development but human development. This needs to be extended by recognizing, in line with Marx, that the real question is one of sustainable human development, explicitly addressing the human metabolism with nature through human labor.

Too often the transition to socialism has been approached mechanistically as the mere expansion of the means of production, rather than in terms of the development of human social relations and needs. In the system that emerged in the Soviet Union the indispensable tool of planning was misdirected to production for production's sake, losing sight of genuine human needs, and eventually gave rise to a new class structure. The detailed division of labor, introduced by capitalism, was retained under this system and extended in the interest of higher productivity. In this type of society, as Che critically observed, "the period of the building of socialism...is characterized by the extinction of the individual for the sake of the state."

The revolutionary character of Latin American socialism today derives its strength from an acute recognition of the negative (as well as some positive) lessons of the Soviet experience, partly through an understanding of the problem raised by Che: the need to develop socialist humanity. Further, the Bolivarian vision proclaimed by Chávez has its own deep roots of inspiration drawing on an older pre-Marxian socialism. Thus it was Simon Bolívar's teacher Simón Rodríguez who wrote in 1847: "The division of labour in the production of goods only serves to brutalize the workforce. If to produce cheap and excellent nail scissors, we have to reduce the workers to machines, we would do better to cut our finger nails with our teeth." Indeed, what we most admire today with regard to Bolívar's own principles is his uncompromising insistence that equality is "the law of laws."

The same commitment to the egalitarian, universal development of humanity was fundamental to Marx. The evolution of the society of associated producers was to be synonymous with the positive transcendence of human alienation. The goal was a many-sided human development. Just as "all history is nothing but a continuous transformation of human nature," so "the *cultivation* of the five senses is the work of all previous history." Socialism thus appears as the "complete emancipation of the senses," of human sensuous capacities, and their wide-ranging development. "Communism, as fully developed naturalism," Marx wrote, "equals humanism, and as fully developed humanism equals naturalism."

The contrast between this revolutionary, humanistic-naturalistic vision and today's dominant mechanical-exploitative reality could not be starker. We find ourselves in a period of imperialist development that is potentially the most dangerous in all of history. There are two ways in which life on the planet as we know it can be destroyed—either instantaneously through global nuclear holocaust, or in a matter of a few generations by climate change and other manifestations of environmental destruction. Nuclear weapons continue to proliferate in an atmosphere of global insecurity promoted by the world's greatest power. War is currently being waged in the Middle East over geopolitical control of the world's oil at the same time that carbon emissions from fossil fuels and other forms of industrial production are generating global warming. Biofuels offered up today as a major alternative to pending world oil shortages are destined only to enlarge world hunger. Water resources are being monopolized by global corporations. Human needs are everywhere being denied: either in the form of extreme deprivation for a majority of the population of the world, or, in the richer countries, in the form of the most intensive self-estrangement conceivable, extending beyond production to a managed consumption, enforcing life-long dependence on alienating wage labor. More and more life is debased in a welter of artificial wants dissociated from genuine needs.

All of this is altering the ways in which we think about the transition from capitalism to socialism. Socialism has always been understood as a society aimed at reversing the relations of exploitation of capitalism and removing the manifold social evils to which these relations have given rise. This requires the abolition of private property in the means of production, a high degree of equality in all things, replacement of the blind forces of the market by planning by the associated producers in accordance with genuine social needs, and the elimination to whatever extent possible of invidious distinctions associated with the division of town and country, mental and manual labor, race divisions, gender divisions, etc. Yet, the root problem of socialism goes much deeper. The transition to socialism is possible only through a revolutionizing practice that *revolutionizes human beings themselves*. The only way to accomplish this is by altering our human metabolism with nature, along with our human-social relations, transcending both the alienation of nature and of humanity. Marx, like Hegel, was fond of quoting Terence's famous statement "Nothing human is alien to me." Now it is clear that we must deepen and extend this to: *Nothing of this earth is alien to me.*

Mainstream environmentalists seek to solve ecological problems almost exclusively through three mechanical strategies: (1) technological bullets, (2) extending the market to all aspects of nature, and (3) creating what are intended as mere islands of preservation in a world of almost universal exploitation and destruction of natural habitats. In contrast, a minority of critical human ecologists have come to understand the need to change our fundamental social relations. Some of the best, most concerned ecologists, searching for concrete models of change, have thus come to focus on those states (or regions) that are both ecological and socialistic (in the sense of relying to a considerable extent on social planning rather than market forces) in orientation. Thus Cuba, Curitiba and Porto Alegre in Brazil, and Kerala in India, are singled out as the leading lights of ecological transformation by some of the most committed environmentalists, such as Bill McKibben, best known as the author of *The End of Nature*. More recently Venezuela has been using its surplus from oil to transform its society in the direction of sustainable human development, thereby laying the foundation for a greening of its production. Although there are contradictions to what has been called Venezuelan "petro socialism," the fact that an oil-generated surplus is being dedicated to genuine social transformation rather than feeding into the proverbial "curse of oil" makes Venezuela unique.

Of course there are powerful environmental movements within the center of the system as well to which we might look for hope. But severed from strong socialist movements and a revolutionary situation they have been constrained much more by a perceived need to adapt to the dominant accumulation system, thereby drastically undermining the ecological struggle. Hence, revolutionary strategies and movements with regard to ecology and society are world-historical forces at present largely in the periphery, in the weak links and breakaways from the capitalist system.

I can only point to a few essential aspects of this radical process of ecological change as manifested in areas of the global South. In Cuba the goal of human development that Che advanced is taking on a new form through what is widely regarded as "the greening of Cuba." This is evident in the emergence of the most revolutionary experiment in agroecology on earth, and the related changes in health, science, and education. As McKibben states, "Cubans have created what may be the world's largest working model of a semisustainable agriculture, one that relies far less than the rest of the world does on oil, on chemicals, on shipping vast quantities of food back and forth....Cuba has thousands of *organopónicos*—urban gardens—more than two hundred in the Havana area alone." Indeed, according to the World Wildlife Fund's *Living Planet Report,* "Cuba alone" in the entire world has achieved a high level of human development, with a human development index greater than 0.8, while also having a per capita ecological footprint below the world's average.

This ecological transformation is deeply rooted in the Cuban revolution rather than, as frequently said, simply a forced response in the Special Period following the fall of the Soviet Union. Already in the 1970s Carlos Rafael Rodriguez, one of the founders of Cuban ecology, had introduced arguments for "integral development, laying the groundwork"—as ecologist Richard Levins points out—for "harmonious development of the economy and social relations with nature." This was followed by the gradual flowering of ecological thought in Cuba in the 1980s. The Special Period, Levins explains, simply allowed the "ecologists by conviction" who had emerged through the internal development of Cuban science and society to recruit the "ecologists by necessity," turning many of them too into ecologists by conviction.

Venezuela under Chávez has not only advanced revolutionary new social relations with the growth of Bolivarian circles, community councils, and increased worker control of factories, but has introduced some crucial initiatives with regard to what István Mészáros has called a new "socialist time accountancy" in the production and exchange of goods. In the new Bolivarian Alternative for the Americas (ALBA), the emphasis is on *communal exchange*, the exchange of activities rather than exchange values. Instead of allowing the market to establish the priorities of the entire economy, planning is being introduced to redistribute resources and capacities to those most in need and to the majority of the populace. The goal here is to address the most pressing individual and collective requirements of the society related in particular to physiological needs and hence raising directly the question of the human relation to nature. This is the absolute precondition of the creation of a sustainable society. In the countryside preliminary attempts have also been made to green Venezuelan agriculture.

In Bolivia the rise of a socialist current (though embattled at present) embedded in the needs of indigenous peoples and the control of basic resources such as water and hydrocarbons offers hope of another kind of development. The cities of Curitiba and Porto Alegre in Brazil point to the possibility of more radical forms of management of urban space and transportation. Curitiba, in McKibbens's words, "is as much an example for the sprawling, decaying cities of the first world as for the crowded, booming cities of the Third World." Kerala in India has taught us that a poor state or region, if animated by genuine socialist planning, can go a long way toward unleashing human potentials in education, health care, and basic environmental conditions. In Kerala, McKibben observes, "the Left has embarked on a series of 'new democratic initiatives' that come as close as anything on the planet to actually incarnating 'sustainable development.'"

To be sure, these are mainly islands of hope at present. They constitute fragile new experiments in social relations and in the human metabolism with nature. They are still subject to the class and imperial war imposed from above by the larger system. The planet as a whole remains firmly in the grip of capital and its world alienation. Everywhere we see manifestations of a metabolic rift, now extended to the biospheric level.

It follows that there is little real prospect for the needed global ecological revolution unless these attempts to revolutionize social relations in the struggle for a just and sustainable society, now emerging in the periphery, are somehow mirrored in movements for ecological and social revolution in the advanced capitalist world. It is only through fundamental change at the center of the system, from which the pressure on the planet principally emanates, that there is any genuine possibility of avoiding ultimate ecological destruction.

For some this may seem to be an impossible goal. Nevertheless, it is important to recognize that there is now an ecology as well as a political economy of revolutionary change. The emergence in our time of sustainable human development in various revolutionary interstices within the global periphery could mark the beginning of a universal revolt against both world alienation and human self-estrangement. Such a revolt if consistent could have only one objective: the creation of a society of associated producers rationally regulating their metabolic relation to nature, and doing so not only in accordance with their own needs but also those of future generations and life as a whole. Today the transition to socialism and the transition to an ecological society are one.

10

The Cyclic Development of the World Capitalist Economy

THE PROBLEM

More than 150 years ago a crisis of overproduction developed for the first time within the young capitalism, seemingly endowed with unlimited forces, that largely determined the subsequent course of the economic evolution of capitalist society. Its epicentre was England, then the main industrial power of capitalism, which had already created a world market and was laying claim to world domination.

Isolated phenomena of the cyclic development of capitalist production had already occurred in separate industries even before the first industrial overproduction crisis of 1825. But it was in the second quarter of the nineteenth century that the cyclicity of capitalism's economic growth began to take on a general international character under the influence of the interconnections of the world market. Marx, attaching paramount importance to this fact, wrote:

> The commercial crises of the nineteenth century, and in particular the great crises of 1825 and 1836 [were] no longer a matter of single economic phenomena but of big storms on the world market, in which the antagonism of all elements in the bourgeois process of production explodes.

From then on the world capitalist system, developing intensively in breadth and depth, began regularly to suffer overproduction crises. In the wake of a cyclic upswing there inevitably set in a time of an absolute fall in or marked slowing of the growth rales of production, international trade, and business in general. The crisis phase was succeeded by a depression, recovery, and a new upswing.

Marx, pointing out the principal historical premises of the origin of the capitalist world cycle, stressed that it was only the period when machine industry, having sent down deep enough roots, exercised a preponderant influence on the whole national production; when, thanks to it, foreign commerce began to take precedence over home trade; when the world market successively annexed vast areas in the New World, Asia, and Australia; and finally when the industrial nations entering the lists had become sufficiently numerous; it was only from that period that the recurring cycles whose successive phases cover years, and which always end in a general crisis, the end of one cycle and the starting point of another.

Study of the laws causing periodic slumps in production within the whole capitalist system and in its component parts, became a cardinal task of Marxist political economy that has retained its importance down to the present.

Quite naturally capitalist economics in turn has paid increasing attention to this, for it, extremely important problem. Each current world economic cycle is analysed in detail, numerous facts are brought out leading to a lowering or a raising of growth rates in one country or another, and in separate industries, but, while performing a certain social job, capitalist economists as a rule willy-nilly try however they can to skirt around those prime causes of capitalism's cyclic development that disclose its historically transitory character.

Their starting point is that each economic crisis, taken separately, is the consequence of certain, even numerous, but wholly removable disturbances of the capitalist reproduction process caused in the main by the concrete market situation at any particular moment. Apart from its purely practical purposes, the work of these writers at the same time pursues a specific ideological aim, i.e. of refuting one of the most important conclusions of Marxist-Leninist political economy, which is that the main contradiction of capitalism manifests itself with irresistible force in cyclic development, and that 'so long as the system lasts', successive cyclic booms and slumps 'must be borne with, like the natural changes of the seasons'.

How many times already has capitalist political economy tried to proclaim the advent of an era of crisis-free development and to give decisive battle to Marxism-Leninism in that connection. Efforts of that kind were intensified after World War II, given economic competition of the two world social systems. The specific features of the development of the world capitalist economy after the war, which accelerated the general growth rates of the productive forces compared with the interwar period, and a definite modification of the duration and depth of the phases of the cycle in the leading capitalist countries created fertile soil for reviving the idea, in the 50s and 60s, that there was a real possibility of capitalism's achieving crisis-free, 'harmonious' growth of production.

A broad propaganda campaign usually develops during upswings around the theories underlying this idea. During the next crisis phase capitalist economists again and again return to the thesis that it is to be explained solely by chance temporary 'difficulties' of one sort or another in the reproduction process having no inherent connection with one another (whose removal would subsequently enable the capitalist economy finally to rid itself of the negative consequences of cyclic development). In any case capitalist writers try to refute, or at least to throw doubt on, the general conclusion of Marxist theory that the business cycle is constitutionally inherent not only in national economies taken separately but also and chiefly in capitalism as a world socio-economic formation.

At the same time a striving to avoid even the very posing of the most acute social problems of the cyclic character of capitalism's development is beginning to be typical of general theoretical work. Characteristically, the commonest theories of recent decades of capitalism's transformation under the impact of the present scientific and technical revolution into some sort of new social system—Bell's ' postindustrial society', Galbraith's 'new industrial society' or 'tcchnostructure', Brzezinski's 'technctronic society', Dahrendorf's 'post-capitalist society', etc.—either ignore tbe principal problems of cyclic booms and crises altogether in practice, or push them far into the background.

Karl Marx, having subjected the socio-economic foundation of capitalism to thorough scientific analysis, for the first time brought out the causal connection between the movement of the world cycle (and the cyclic development of its components) and the deep antagonisms of this mode of production. A sharp polemic develops in capitalist political economy and sociology around this Marxist tenet during each successive economic crisis. Many of their loading spokesmen, moreover, while not denying that economic cycles and crises used to be inherent in capitalism, base themselves on its having passed, since the war, to an allegedly fundamentally different stage of development, in which its former antagonisms are losing their sharpness and may be successfully overcome by state capitalist programming and control.

In this connection a version has begun to be intensively disseminated in the Western literature that the specific features of the postwar'dynamics of capitalist production (and to a decisive degree of industrial production) allegedly do not provide grounds for concluding that the patterns of its cyclic development disclosed by Marxist political economy continue'to operate. It is often stated, moreover, that if the movement of industrial cycles and crises can be observed within separate national economies, this cyclicity lias either 183 disappeared or begun to do so on a world scale since the war.

Hence optimistic conclusions are drawn about capitalism's growing over into some phase of 'harmonious' development. The three economic slumps suffered by world capitalism in the 70s and early 80s, it is true, led to a marked fall in this optimism. They again graphically demonstrated that the cyclic upheavals underlying crises of industrial production are irrevocable features of the postwar development of the whole socio-economic system of capitalism as previously.

It lias become very obvious, moreover, that in spite of the present-day historical situation's extremely essential differences from that established in the world economy at the start of the 30s, the menace of a repetition of all-embracing cyclic upheavals on the scale of the 1929–33 crisis has by no means disappeared. Furthermore, as the deep-seated contradictions of capitalism are aggravated, with timo this menace continues to grow. That fact has been more and more widely recognised in capitalist economic literature since the mid708.

THE POSTWAR INDUSTRIAL CYCLE

The scale and periodicity of the business cycle are primarily characterised, in their most general forms, by indicators of the dynamics of gross production. These characteristics are quite justified, since the fluctuations of the cycle embrace more or less all the main spheres of the business activity of capitalist society. By no means all the sectors of social production involved in creating the gross product, however, play the same role in the shaping of the mechanism of the capitalist cycle. Most of them (agriculture, the services sphere, transport arid communications, etc.) either reflect or only indirectly affect the process, which is underlain (as MarxistLeninist political economy has established) by cyclic fluctuations of the decisive sphere of material production, i.e. industry.

As we have already noted the postwar course of the general development of capitalist economy, its cyclic development included, does in fact have essential features that are displayed in a marked intensification of the urievenness and spasmodic growth of its constituent parts. These features in turn have greatly complicated elucidation of the interaction of the cyclic movement of the individual national economies' production. Still, this interaction of the national cycles within a single world cycle is clearly traceable in the postwar period.

It follows from the table, first of all, that the capitalist conception of crisis-free, 'harmonious' development of tlie economy of postwar capitalism is by no means built on facts. At the same time the table confirms tbe truth and validity of the conclusion of the 1969 International Meeting of Communist and Workers' Parties that state-monopoly regulation, exercised in forms and on a scale which moot the interests of monopoly capital and arc aimed at preserving its rule, is unablo to control the spontaneous forces of the capitalist market. Practically no capitalist stale has been able to avoid considerable cyclic fluctuations arid slumps in its economy; in some countries periods of rapid industrial growth alternate with periods in which there is a slowdown and often a drop in production.

But state intervention in economic affairs was unable to eliminate the causes of this 187 cyclicity. At best the disease was postponed. At the same time the counter-crisis measures implemented by governments began to affect the specific postwar features of the movement of the separate phases of their national cycles to one degree or another, and consequently the world industrial cycle as well. Among the other factors operating on the postwar features of the cycle the commencing process of accelerated growth of the economies of developing countries occupied a notable place. In the vast majority of the latter, as we kno\v, counter-cyclic regulation was not practised; it must be noted, however, that their annual growth rates of industrial production have been rather higher and stabler in recent decades than in capitalism's industrial centres.

In the years considered a quite consistent succession of periods of a rise and fall in the aggregate growth rates of the separate countries' industrial production is traceable through the ceaseless short-term fluctuations of their economies. Each fluctuation has its own peculiar features, but there is nothing unusual in that. There has not been a single world industrial cycle during capitalism's whole development that fully repeated the preceding one and did not reflect specific features of the current state of the economy of some one of the main groups of countries and of the internationalising of their social production. The problem of the interaction of the world cycle and the cyclic dynamics of production in the leading capitalist countries inevitably arises when we are studying these features.

The two substantial declines in the general growth rales of industrial production in the 50s were the result of a quite significant economic crises in the periods concerned in the economic situation in almost all the developed countries. There was also a certain slowing of growth in the same periods in the whole group of developing countries. This process evolved on the background of an intensification of the asynchronous character of the cyclic movement of the economies of the various regions and countries connected with the long-term consequences of World War II. While there was an almost universal slowing of industrial growth rates in the early 50s, the most marked decline was experienced by Western Europe, where there was an absolute fall in production in several countries (Great Britain, France, Belgium, etc.).

Later the economies of most of Hie West European countries, Japan, and a considerable number of developing countries picked up quite rapidly and entered on an upswing. The North American countries, however, reached this upswing much later. From the middle of 1953 there were again deep crisis phenomena there that continued for the greater part of the next year, which led to a fall of industrial production by 5.2 per cent in the USA and 1.7 per cent in Canada. That ultimately conditioned the zero total growth of production in that year, since the USA was producing more than half of the total industrial product.

During the futher movement of the world cycle, in the course of which the peak growth rales were reached in 1955 (10.5 per cent), the temporary upswing phase prepared the ground for the next crisis slump, signs of which began to show clearly at the end of 1957. The low point of this cycle was reached in 1958, when industrial production fell for the first time since the war in absolute terms, by 2.5 per cent. The crisis swept all the industrial countries without exception almost simultaneously, but with varying force. The epicentre was the USA where the annual volume of industrial production declined by more than 6 per cent. Most of the countries of Western Europe, Canada, and more dynamically developing Japan (whose annual increment of industrial output fell from almost 25 per cent in 1950 to 15 per cent in 1957 and 1 per cent in 1958), were drawn into it.

A characteristic of the 1957–58 crisis was that many of the countries of the primary commodity periphery had begun to 189 experience serious economic difficulties even at the end of the preceding upswing, although their industrial growth rates as a whole did not substantially decline. At the same time the leading sectors of their economies, especially the raw material industries oriented on export, were in an extremely grave position. All that went hand in hand with a marked weakening of their position on the capitalist market. Separate countries, and groups of countries, did not experience the consequences of this crisis of the late 50s to an equal extent, and the crisis phase was not protracted. World industrial production soon again reached quite high growth rates (in 1959 growth was more than 9 per cent).

The capitalist economy of the 60s was characterised by a further intensification of the unevenncss of the growth of its various parts, which could not help affecting the dynamics of its cyclic development. In several industrialised countries and many developing ones the tendency noted in the 50s toward a certain acceleration of industrial growth rates was maintained, which encouraged a strengthening of the asynchronous character of their movement in various countries and groupings. The amplitude of the cyclic fluctuations of world industrial development indices was less marked in the 60s, on an average, than in the preceding decade. The consequences of the major recessions in some countries were compensated one way or another on the whole by higher growth rates in others. There was a marked slowing of production rates in 1961, for instance, in the USA, Great Britain, France, West Germany, the Netherlands and certain other countries, associated with features of their cyclic movement, and bearing the character of a partial overproduction crisis. But because of the quite high growth rates in several other countries, including Japan (20 per cent), Italy (9 per cent), and the developing countries (7.5 per cent), the average annual increment of the industrial product was 3.5 per cent in 1961 for the whole non-socialist world, against 6 per cent in the preceding year.

In the following years the industry of France (1905), Great Britain (1962, I960), Italy (196/j),'Japan (19(15), Belgium (19GB), West Germany (1967), and certain oilier counlries, allernalcly experienced crisis phenomena, and there was a marked lowering of industrial growth in lhe developing countries as a group (from 13 per cent in 19(50 lo 4 per cent in 1962). Still, the annual growth of world capitalist industry was comparatively high and stable, not falling below 5 per cent, and being (5.9 per cent on an average for the period 1902–60.

It was in those years that views began to be very widely disseminated in capitalist economic literature that the Marxist theory of the cyclic dovelopmenl of the capitalist economy was being refuted by the real course of events, since crisis recessions in tlic economies of individual capitalist countries had now allegedly begun to be local only and had uo fundamental influence on the development of the world system.

The period favourable for broad propagation of these views was not, however, very long. In 1967 there was a substantial lowering of industrial development rales on a world scale (lo 2.5 per cenl), which showed that there was again a strengthening of the synchronous character of their development in most of the industrial centres, although a relatively high stale of business was maintained in some developed counlries (e.g. Japan and Ilaly), and in most of the developing countries.

The brief boom thai began laler, which was accompanied by an increase in the world production of 8.5 and 7.5 per cent in 1908 and 1909 respectively, was in fact the forerunner of a significant fall in growth rates at the very beginning of the 1970s. As in the preceding crisis phases the centre was in Ihe USA; at the same lime the crisis manifesled itself painfully in the economies of such major countries as West Germany, Great Britain, and Italy, and to a considerable exlenl also in Japan, Belgium, and olher counlries. The national cycles were more synchronous, though they remained far from identical, and during their evolution there was no sleep curtailment of the movement of production as a whole in the developing countries and certain individual developed counlries (France and llu; Netherlands).

This intensification of the unevenness in Ihc time, scale, and frequency of the sequence of the individual phases of Ihe industrial cycle in various countries, which was averaged out and smoothed out in Ihe summary international indicators, led to the aggregate industrial production of world capitalism not falling in absolute terms between 1959 and 197/I, i.e. for a space of 10 years. True, the gap between the highest and lowest annual growth rales was very considerable. In the late 60s and early 70s, for instance, it atlained Ihe value of a faclor of more than three (the growth of industrial produclion of the whole non-socialist world was 8.5 per cenl in 1968 and 2.5 per cent in 1971).

Was this asynchronousness of the poslwar industrial growth rates of individual countries a qualitatively new thing in the world capitalist economy? This point is becoming especially acute and pressing, for another posing of the problem follows logically from it, viz., was there a single orld cycle in the first postwar decades or did the main trends of development of capitalist society's productive power lead lo its disintegration more and more into mutually isolated national cycles?

Synchronism and Asynchronism in the Evolution of the World Cycle

The non-coincidence in the time, duration, and intensity of the different phases of the cycle, and in the depth of the crisis in the individual national economies, of Ihe 50s and 60s was not at all unique in the history of the world economy. On the contrary, the whole experience of the evolution of the capitalist mode of production indicates that such noncoincidence was frequently a regular form of the functioning of the world cycle.

Although the inner unity of the national cycles, linked togelher by an appropriale syslem of international division of labour, is expressed in the dynamics of the movement of production in the world capitalist economy, that in no way signifies that they cannot manifest themselves at different times, in different months and even years, within the context of any one world cycle. Furthermore, it inevitably remains quite probable in such a complicated, many-sided process that there will be a non-coincidence of the cyclic movement of production in separate countries, and that there will often be considerable differences in the aggregate forms of the display of this movement from the specific forms of its dynamics within the separate national economies.

When we turn, in this connection, to indisputable historical facts, they prove that the principal capitalist countries repeatedly entered the crisis phases of the business cycle at different times in the past. Over the period from 1857 to 1933 the capitalist system experienced, we know, nine world overproduction crises. In only two of them (those of 1857 and 1907) did the crisis develop in all the main countries in the same calendar year.

That, however, in no way means that the outlines of the world cycle became more and more blurred during capitalism's conversion into an all-embracing system, or that it had already begun to disintegrate into national economies at the end of the nineteenth century. Its whole history indicates that a tendency to broaden the objective conditions of a single world cycle has predominated.

It also does not follow from analysis of the postwar cycles that there was only a tendency then toward asynchronism, or that Hie world cycle had become more and more ' disjointed'.

An opposite tendency was operating simultaneously that was above all a consequence of the whole growing internationalisation of social production and led objectively to a stronger convergence of the course of the cyclic 193 development of the separate countries and principal industries. One or the other of these trends is manifested more prominently under the impact of a whole host of concrete circumstances. And there are adequate grounds, for all that, for thinking that an obvious connection between the periods of cyclic rise and fall in production growth rates predominated in the main industries of the capitalist world as well as in most of its countries.

By Ihe middle of the 60s relative high indices of industrial development again began to be typical of the majority of capitalist and developing countries, which again declined in many of them from the beginning of the second half of the decade. That led in 1967 to a substantial fall in the world growth rates of capitalist industry, after which a brief cyclic boom began again in the overwhelming majority of capitalist countries.

At the very beginning of the 70s the economies of almost all these countries experienced a fall in growth rates at the same time, which led to an absolute fall of production in some of them. But a little later the economy began to pick up in all the industrialised countries in 1971, and their production of industrial goods increased somewhat on an average. The peak of the upswing in this cycle was also reached almost simultaneously in the overwhelming number of capitalist and developing countries. As a result capitalist industrial production rose by 7 per cent on a world scale in 1972 and by 9 per cent in 1973.

Thus, there was a definite strengthening at the beginning of the 70s of the tendency toward a synchronising of the country-by-country movement of industrial development, which showed particularly clearly in the middle of the decade when the industrially developed countries began, one after the other, to enter the longest arid deepest postwar world economic crisis. This trend was undoubtedly strengthened because the crisis developed in a period of acute aggravation 194 of the deep conIradiclions of capitalism's whole system of world economic relations. Total industrial output consequently did not increase in its economic centres in 1974 and fell hy nearly 7 per cent in 1075. At the beginning of the 80s industrial production fell markedly in the industrial centres of capitalism compared with the preceding decade. For most countries the recovery from this crisis has the temporary character typical of recent world business cycles.

In 1076 the total volume of industrial production of capitalist and developing countries was more than 8 per cent above the 1975 level; later, however, as in several preceding cycles, there was a slowing of growth rates primarily associated in the developed countries with the sluggish upswing in the leading West European countries, and in the developing countries with the comparatively slow rates of many Latin American and African countries. At the same time North America and several developing countries in Asia, especially in the Middle East, showed quite considerable growth. The growth rates of industrial production in North America, for instance, were 50 per cent higher than in Western Europe in 1976–79, and in the developing countries as a whole somewhat higher than in Latin America.

There was thus a certain strengthening of the asynchronism of the development of the separate groups of countries with this phase of the world capitalist cycle. This asynchronism markedly weakened after the USA, Great Britain, Canada, France, Belgium, Sweden, and most other capitalist countries entered the next loop of economic slump at the turn of the decade to the 80s. In 1980, compared with the preceding year, the capitalist countries' general volume of industrial production declined by roughly 0.5 per cent. At the same time there was a steep fall in industrial growth rates in all the main regions of the developing world. Annual index numbers cannot, however, be considered adequate for a comprehensive description of the capitalist economy's cyclic movement.

They only let us distinguish the general trends of this movement since they average out and smooth over the course of the cycle within shorter intervals. We therefore need more detailed index numbers in order to study problems of the simultaneous character and lack of synchronisation of the cyclic fluctuations of the different parts of the world economy and of the depth of the gap between I ho highest and lowest points of each cycle. In (hat respect it is very important to analyse the quarterly figures, since that lets us get a clearer idea of the build-up of the trends of economic development during capitalism's feverish fluctuations by separate quarters. A graphic illustration of this is given by the quarterly movement of the index numbers of industrial production in seveial of the main regions at the end of the 60s and in the early 70s. Above all, they indicate that there are quite sizable variations in the rates of decline and growth of production within one and the same phase of a cycle. In North America, for instance, where there was a substantial drop in industrial production in 1970–71, these rates did not decline steadily.

The processes considered were displayed with special force and clarity during the world economic crisis of the mid-70s (the biggest since the war), and later again in the crisis of the early 80s. The declines in production touched bottom respectively in the third quarter of 1975 and 1980. In the first case the level of industrial production of the developed capitalist countries was nearly 11 per cent below the quarterly high of the preceding cyclic upswing, and in the second case 8.1 per cent below.

Analysis of the quarterly movement of world capitalist industrial production, and of its main components, gives grounds for stating that there was certainly an absolute decline in the volume of production in the 70s and the early 80s in the overwhelming majority of countries between the peak and bottom quarters of each of Ihe 18 periods of world industrial recession noted in the table.

One can say, as a result, that the quarterly changes in production, for all their dependence on seasonal and oilier current factors, make it possible to gi\e a clear picture of the length, scale, and course of the world capitalist cycle. At the same time we must not forget that the quarterly figures by themselves, out of context of the annual dynamics of the movement of production, cannot be considered the decisive instrument or the principal statistical source of information for studying the long-term trends of world cyclic fluctuations. For in periods of comparatively high business activity one repeatedly meets cases of a sharp slowing or even drop in production in separate quarters, associated with seasonal or other short-term fluctuations of the market situation. Study of the relations of the various sectors and spheres of world production in their cyclic development must therefore be based first of all on analysis of the annual indicators, which help us bring out the resultant trend of the long-term dynamics of the fluctuations of production and the market within a single world business cycle.

The Capitalist Market and World Cycle

The idea that the course of world capitalism's cyclic development reflects some mechanical aggregate of the movements of national cycles is as illegitimate as the proposition that the cyclic growth of capitalist product ion within the separate, nationally isolated economies is based on their internal factors of development. The national economies of individual countries are constitutionally linked into a single economic organism by the system of international division of labour on the world capitalist market.

Karl Marx repeatedly stressed that the world capitalist economy's cyclic movement found its general expression in regularly recurring crisis fluctuations of the world market. 199 The latter, being the final result of the interlocked cyclic development of the national economies, embrace the most important aspects of the reproduction process of the separate countries.

The world trade crises must be regarded as the real concentration and forcible adjustment of all the contradictions of bourgeois economy. The individual factors, which are condensed in these crises, must therefore emerge and must bo described in each sphere of the bourgeois economy and the further we advance in our examination of the latter, the more aspects of this conflict must be traced on the one hand, and on the other hand it must be shown that its more abstract forms arc recurring and are contained in the more concrete forms.

This conclusion is logically linked with other very important tenets of Marxist-Leninist political economy that bring out the patterns of the functioning of the capitalist mode of production as a world system all parts of which are united into a single whole through the international division of labour.

Already in the middle of the nineteenth century Marx had noted that large-scale industry was closely dependent on the world market and the international division of labour. Developing that idea ho subsequently showed that the world market at onco is the precondition and the result of capitalist production.

Since these theoretical points were formulated, the internationalising of social production by capital has led to an immense strengthening of international business connections on the world market. In spite of repeated crises, international trade rose by a factor of five in the second half of the nineteenth century. From 1900 to 1938 it doubled again. After World War II it grew particularly rapidly. The physical volume of foreign trade within the bounds of postwar world capitalism, much reduced by socialism, had increased more than sevenfold on the whole by 1978 compared with 1938, and trade in industrial items more than tenfold.

The graph indicates that a definite cyclic trend of development of a long-term character made itself felt, as before, in the postwar period, through countless fluctuations of the elements of the world market. After each upswing of the growth rates of trade, there was invariably a recession, followed by a new upswing fraught with another recession, and so on. In the period under review several such cycles can be traced in trade as well as in industry.

Thus, the very substantial fall in the growth of world trade in the early 50s, which led to an absolute reduction in its volume in 1952, was later succeeded by an upswing that reached a peak in 1955. The next cycle began with an even more marked slump in 1957–58. Later the outlines of this type of cycle were not so marked, especially in its lower phases, but in those years, too, there were several cycles. Their troughs occurred in 1961, 1967, 1971, and 1975. In other words, for all the differences, the length of each of them averaged four to six years and was noticeably shorter than the periodicity of the market's cyclic fluctuations in preceding stages of its evolution.

The following features in the analysis of the dynamics of the growth rates of international trade since the war catch one's attention. Whereas the trough phases in the 50s were marked by an absolute fall in the physical volume of trade, such was not the case in the 60s. The trough was relatively shallower and did not sink below 4.2 per cent in relation to the preceding year. Hence the question naturally arises whether this feature of the market, characteristic of the 60s and early 70s, was a quite new phenomenon, i.e. whether two qualitatively distinct stages must be distinguished when analysing its growth in the first postwar decades. For in the 50s the crisis phases of the world cycle ended in an absolute reduction in the scale of international trade, while in the 60s up to the mid-70s they were expressed in a relative fall in its growth rates. Only in 1975, for the first time since 1958, was there a marked contraction of the world capitalist market.

Here the approach to evaluating the ratio of the indicators that characterise the boom and crisis phases during 202 the modern capitalist economy'scydicdevelopment inevitably conies to the fore. The point is the following: is it legitimate to include years in the crisis phase when there is a substantial slowing of the growth rates of the productive forces, while such very important indicators as the volume of world production and international trade do not fall below the level of the preceding years of cyclic upswing? Another posing of the problem also follows from that: is there a sharp dividing line between the cyclic dynamics of the evolution of world capitalist economy in the past and in the postwar decades we are considering? The approach to clarification of this point is linked, of course, with the need to consider the corresponding facts of the previous history of capitalism's cyclic development.

The view that there is an inherent crisis phase in which there is necessarily a deep, protracted fall of industrial production and international trade used to be quite commonly held in the Soviet literature. The world crisis of 1929–33, in the course of which the physical volume of international trade fell by almost two-thirds and world industrial production by around a fifth, had considerable influence on the forming of this point of view. But, as the experience of the development of world capitalist economy has shown, that crisis was unique. The slumps of even the 'classical' world cycles of the turn of the century, for which there are summary statistics, were by no means always, marked by an absolute reduction of the indicators of the system's international business activity.

Thus, in the 1882 crisis, world trade grew by around 4.5 per cent, but compared with the preceding boom period, this was a fall of 50 per cent (growth was 9.3 per cent in 1880). The next year, with a certain absolute fall in the volume of industrial production, the growth rates of trade fell significantly, though remaining above the 1882 level in physical volume. The cyclic crisis of the early 1890s was not accompanied with an absolute fall of world trade. In 1890 its growth was around 3 per cent against 7 per cent in the preceding year.

In spite of the subsequent decline in the .scale of world trade, its lowest absolute level in the 1890s cycle was above the corresponding high of the preceding cycle. In the 1900 crisis world industrial production increased on the whole by more than 5 per cent and. the commerce of the principal 203 trading countries by (i per cent. According to the UN statistics, the next year the index of the physical volume of international trade remained at the 1900 level, while the index of world production of manufactures even rose somewhat.

Estimates based on the statistical' publications of the League of Nations, provide evidence of the dynamics of manufactnring industry during the successive world economic crises of the turn of the century. For all the conditional character of the summary indices presented, they make it possible for all that to make certain generalisations of a comparative character.

They allow us to conclude that the course of the world market's cyclic development, when the annual indices of the physical volume of international trade did not fall in absoule terms in the lowest phases of several postwar cycles, or fell very temporarily, is not a phenomenon that emerged then. Such cases have heen repeatedly met in the history of capitalism.

The same can be said of the spasmodic character of the dynamics of industrial growth on the scale of capitalist economy as a whole. During its long evolution there also used to he world cycles that were characterised hy substantial, but only relative, declines in the growth rates of world trade and industrial production compared with the preceding upturns.

It would hardly be justified, in that connection, to consider that the decisive criterion of the onset of a world crisis was also, in the earlier stages of capitalism, an absolute and prolonged decline in the main annual indicators compared with the level already reached in preceding years. The periodicity of the cyclic fluctuations, and the scale, depth, and length of world economic crises, are determined by the main patterns of this mode of production in its most general form. In real life, the concrete course of each of them depends on the interaction of a host of historical 'accidents' and the current situation in the development of the economies of the separate countries within the system of the international capitalist division of labour.

As E. S. Varga justifiably remarked, each world cycle and crisis takes on a specific character determined by a host of very different factors of both an economic and an extraeconomic order. Moreover, not only does each world crisis have its own special, typical features but development also proceeds unevenly in the separate countries; and within them the separate industries develop extremely unevenly, there being fast growing 'new' industries and alongside them 'old' ones displaying a tendency to decline irrespective of the cyclic curve of reproduction. The correctness of these statements is confirmed by the whole experience of the cyclic development of the world capitalist economy, but that experience does not provide the necessary grounds for recognising as legitimate in all respects the view expressed by Varga in the mid-30s that with the development of capitalism crises cannot be smoothed out but must become more and more universal, deeper, and more and more frequent.

In any case, the summary statistical indices of the rates of postwar development cited above clearly indicate that.

The frequency of world recessions of business activity has increased, but not one of them has yet been as deep or on the scale of the 1929–33 crisis. Furthermore, these crises have been characterised in the main, on the background of the certain acceleration of the growth of the productive forces in most capitalist and developing countries in the 50s and 60s, by a relative decline in the growth of international trade and production, when comparatively deeper crises alternated with shallower ones.

Study of the specific character of each of the postwar world cycles undoubtedly requires us to allow for such concrete historical factors as the various forms of manifestation of the scientific and technical revolution in one industry or another, the steady growth of economic competition between the two opposing social systems, the extension of state-monopoly control in the leading capitalist countries and on a world scale, the socio-economic consequences of the break-up of the colonial system, the intensification of the class struggle in the developed capitalist countries for their workers' vital rights, and so on.

Even the deep cyclic crises of the world capitalist economy accompanied with a drop in the volume of world industrial production and international trade in fact have a temporary character, as they used to, since they are compared only with the preceding boom phase. At the same time the low of each new crisis is higher, as a rule, in absolute terms, than the corresponding level not only of the previous crises but also of the cyclic upswings before it.

This is one of the most essential patterns of the growth of the world capitalist economy. Crises periodically hold hack its development and at the same lime create the objective conditions for the next upturn. The capitalist economy experienced four world economic crises, for instance, between 1880 and 1913; international trade increased over, the same period by approximately 170 per cent, it- annual growth rate being around 3 per cent. The volume of world industrial production increased correspondingly by a factor of four, with an annual growth rate of more than 4 per cent. In the next 35 years (1913 to 1948), which covered two world wars and the most acute cyclic crisis (1929–33) in the history of capitalism, the annual rates of economic growth fell steeply. In the sphere of international trade they were only 0.6 per cent, and in industrial production 2.4 per cent. In the 50s through the 70s the corresponding annual indicators within the contemporary world capitalist economy were approximately 6.8 and 5.0 per cent.

Thus, over a long period of ca, italism's development, an absolute fall in the generalised indicator of the volume of trade and industrial production was by no means always a longterm orientor for evaluating the crisis phase of its world cycle. A relative, but more or less protracted, deep fall in the aggregate growth rates of international trade and industrial production can also serve as an indicator of this. In the same way the transition from the crisis phase to the boom within one world cycle may be determined by the development of a tendency toward a considerable acceleration of economic growth relative to the low point touched in the cycle just completed, rather than by a surpassing in absolute terms of the maximum level of development of the preceding cycle.

To characterise several successive world cycles, we have to establish a system of indicators that would 'et us compare the scale of the fluctuations of the extreme phases of cycles. For that the annual growth rates over a certain period of time are above all of special value, since they make it possible to trace the long-term trends of cyclic movement on a broader scale than quarterly or monthly data.

At various stages these tendencies, of course, acquire very substantial differences. Each of them, normally covering several consecutive cycles, reflects on a longer plane the unevenness and instability of the world capitalist economy in certain conditions of its historical development.

As we have already remarked, the annual growth rates of international trade in the decades before World War I substantially exceeded the corresponding indicators of the interwar period but were much below those of recent decades. Therefore the growth rates of the physical volume of trade at the turn of the century, which were 4 or 5 per cent per annum, could have been evidence of a cyclic boom, while now the same growth may serve as the indicator of the next cyclic downturn. Thus the negative trend manifested in international trade in the mid-60s led to an approximately 50 per cent fall in its physical volume, from 10 per cent in 1964 to around 5 per cent in 1967. A considerable fall occurred in the course of the world cycle of the late 60s and early 70s, when the growth rates of international trade fell by almost half in 1971 compared with 1968.

But perhaps the substantial changes that have occurred in recent decades in the world market have led to its cyclic fluctuations no longer adequately reflecting the cyclic character of the world capitalist economy as a whole, above all of its industrial basis, as happened in Marx's day? To clarify this point let us compare them with the indicators of the postwar development of industrial production and gross product on a world scale. The long-term trends brought out in the graph very obviously indicate the maintenance of a synchronous movement of these indicators, which in their aggregate reflect the regular periodicity of the main phases of the modern world capitalist cycle. Instability, a spasmodic character, and cyclicity are, as before, integral features of world capitalist business, however specific the postwar conditions of development of its productive power.

Industry in the Modern Cycle

The driving force of world capitalism's cyclic development, according to Karl Marx's conclusions, lies in the sphere of industrial production. The indices considered above characterise the overall picture of this development in its most general features. They need further refinement and concretisation, of course, both for the separate industries and the individual countries.

The marked differences between industries on this plane are initially characterised by the summary indices of the cyclic growth of light, and especially of heavy, industry. It is in the manufacturing sectors of heavy industry (which has been taking an increasingly leading place in recent decades compared with light industry) that the modern world industrial cycle is largely generated. And the main impulses of the development of this cycle come quite logically from the developed capitalist countries.

In an analysis of the long-term trends of development of the world business cycle, the fact that periods of boom and downturn of the growth rates of light and heavy industries have converged quite closely is of paramount importance, but the range of the fluctuations between the highs and lows of the cycle in the light industries has not been so considerable as a rule as in the heavy industries, which is a consequence in the main of their slower growth rates. In other words, the features of the latter cyclic movement compared with the former's have led to a certain smoothing out of the acuteness of the cyclic fluctuations reflected in the summary indices of the development of manufacturing as a whole, and consequently of total world industrial production.

In fact, there has been no major branch of manufacturing industry in the postwar period whose development has run counter to the dynamics of the world capitalist cycle c nsidered above. But in some industries, because of their specific conditions, the cyclic fluctuations have not been so distinct and have been smoother. That applies above all to the food and drink industries in which the fluctuations of growth rates have always been relatively shallow in contrast to those, for example, of such light industries as textiles, clothing, and footwear. In the postwar period (with the exception to the crises of the mid-70s and early 80s) annual growth of the 213 production of foodstuffs did not fall below 3.5 per cent, but also did not go higher than 5.5 per cent. Even during these crises, however, there was no absolute decline of production, although the growth curve sagged steeply. Over the postwar period to 1974 inclusive there was no decline in the annual growth rates of one of the most dynamic spheres of light industry, viz., paper and printing. In those years only once, in 1957, was there an absolute decline in the volume of production of the woodworking industries, but during the last two crises there was quite a substantial decline.

There are also branches of heavy industry in which there was only a relative slowing of the growth rates of production in the crisis phases of the postwar world cycles. In the chemical industries, for example, developing very rapidly under the impact of scientific and technical progress, growth rates were not once below 3 per cent, except in 1975, although the cyclic variations in them were quite marked. In the cycle of 1954–58, for example, the gap between the highest and lowest annual levels was fourfold (13 per cent and 3 per cent). Thus, separate groups of industries affect the course and development of the world cycle differently. The basic metal and metal-working industries have a leading place in this respect. Their cyclic fluctuations, combined with corresponding spurts and declines in the growth rates of several light industries (textiles, clothing and footwear) play the most active role in shaping the postwar industrial cycle. They have also primarily determined the depth of its crisis phases.

The four groups of industries reviewed below include, as before, the bulk of the industrial output of the capitalist economy. Their development trends mainly characterise the cyclic course of production of the corresponding light and heavy industries. Despite the quite natural differences in their dynamics, the graph brings out with extraordinary clarity the same temporal synchronism of cyclic fluctuations, with an average periodicity of four to six years, which cannot, of course, be explained at all by patterns in the periodic renewal of their fixed capital. The periods of this renewal, moreover, are essentially specific in each industry and usually have a much longer character linked in the main with the gradual obsolescence and wear and tear of equipment.

The prime causes of the periodicity of the fluctuation of industrial, and generally of economic, cycles noted in the postwar capitalist economy, are undoubtedly governed by a complex set of very different factors. To explain them calls for a special industry-by-industry and country-by-country analysis, but it is clear in any case that this periodicity stems from the regular interconnection objectively established in recent decades between the cyclic variations of the growth of industrial production and the ratio of aggregate supply and demand on the world and national markets, which spontaneously govern the volume of this production. The contracting periodicity over the postwar decades in turn characterises the growing instability of capitalism's economy. Study of this consistent postwar periodicity is essential in order to explain not just the past and current trends of world capitalism's development, but—and this is the main point—the outlook for coming decades.

Can We Forecast World Economic Crises?

It is becoming of paramount importance for analysis of the historical future of capitalism to investigate the longterm outlook of its cyclic development. It is therefore not fortuitous that this study is engaging the close attention of both Marxist and capitalist economists. The ideological struggle between them in this field is developing mainly around the fundamental causes of business cycles, especially around whether or not it is possible to eliminate sharp cyclic fluctuations of the economy within the framework of this mode of production, rather than in the sphere of estimating the scale and duration of already past cycles.

Since the facts of the past speak for themselves, capitalist economists and sociologists always make great efforts to refute the Marxist-Leninist theory of cycles and crises at least when projected into the future. Their aim is to question, by any means whatever, the legitimacy of employing this theory to analyse any subsequent stage of the development of capitalism. Given the economic rivalry of the two world social systems, many present-day capitalist economists have begun to place special hopes on state-monopoly methods of counter-crisis policy. The compilers of President Kennedy's economic report to Congress, attaching decisive importance to these methods, remarked, for example, in 1962 that 'the business cycle does not have the inevitability of the calendar' for the United States. The same idea was also put forward in his next report. But such statements are clearly at .variance with all the facts of the past and with the postwar trends of the economies of the USA [and other capitalist countries.

As we have already remarked, there is a tendency in the Western literature, especially in publications that do not have very great confidence in the traditional postulates of capitalist political economy, to recognise the inevitability, in one way or another, of the cyclic development of the capitalist economy and its regularly alternating slumps and booms. But they combine attempts to examine the objective causes of the cyclicity with a steady striving to explain them each time by concrete spontaneous processes of one sort or another of the economic situation without reference to the general socio-economic patterns of the functioning of the capitalist mode of production.

As Karl Marx pointed out in his day, students of capitalism's cyclic development treat every new crisis as an insulated phenomenon, appearing for the first time in the social horizon, and, therefore, to be accounted for by incidents, movements and agencies altogether peculiar, or presumed to be peculiar, to the one period just elapsed between the penultimate and the ultimate revulsion. If natural philosophers had proceeded by the same puerile method, the world would be taken by surprise on the reappearance even of a comet.

The same methodological approach continues to dominate capitalist economics today. The well-known American economist J. K. Galbraith, for instance, characterising the causes of the 1974–75 crisis in an interview entitled 'Why the Economy Is in a Mess—and What to Do about It', said: The recession was a matter of policy. I would attribute most of it to the use of tight money—-to the use of high interest rates to curb bank lending and thus to reduce inflation. There was, additionally, some overflow effect from the high oil prices.

Another American sociologist, Herman Kahn, Director of the Hudson Institute, declared at the end of 1974:

> My feeling is that we are going through a transition that could make the 1970s a turning point in the history of the world. In other words, we could be on the threshold of a long period of orderly and sustained growth which will solve many of the problems everybody is worrying about today In fact, our studies at Hudson Institute suggest that what we're actually going through now are mostly the growing pains of success and some elementary mismanagement.

In addition, one often finds the idea in the postwar Western literature that capitalism will have to reckon in the future on the inevitability of disruptive consequences from future cyclic crises. The spokesmen of capitalist ruling circles are caused special disquiet by the fact that any ordinary economic crisis may he fraught, in a situation of an orgy of inflationary chaos, mounting instability of world capitalism's monetary and financial system, and of sharp aggravation of its energy problems, etc., with far-reaching social upheavals. They naturally see the prime cause of such a danger, of course, not in the nature of capitalist production itself, or in its unresolved contradictions, but in the probability of individual statesmen pursuing 'shortsighted', 'incompatible' economic policies, or in serious disturbances of the machinery of world economic relations like the energy crisis, etc.

When, for example, the world capitalist economy began, after the brief boom of 1972–73, to slide again into a recession, the British weekly *The Economist* wrote:

> Britain may bo heading for its worst slump since the interwar years, because of an undeliberated repetition of the mistakes of 1970–71 on five times the scale.

About the same time David Rockefeller, chairman of the Chase Manhattan Bank, warned in New York that the energy crisis 'could' precipitate a world depression.

During the 1074–75 world crisis the forecasts of economists in several capitalist countries began to foresee the probability of the onset of a subsequent business crisis.

Such forecasts are very symptomatic, and reflect recognition of the fact that state-monopoly capitalism's many counter-crisis measures do not yield the expected effect, that the ultimately unmanageable chaos of capitalist economy engenders the cyclic character of its development and objectively carries within it a destructive potential of considerable force. For instance, a number of leading capitalist theoreticians of government counter-crisis policy had already been bitterly forced in the 60s to recognise its feebleness.

When attempts are made in the capitalist economic literature to peer into the future movement of the world capitalist cycle, they do not usually pursue the goal of an, in any way long-term, forecast, but are limited to evaluating the immediate prospects, and as a rule are made in conditions when the main economic indicators of several countries have already pointed to the approach of another cyclic crisis.

A very important place is assigned here, moreover, to modern computers. A writer in *The New York Times*, being ironic in this connection about economists' lack of any theoretical developed conception of the cyclic movement of political economy, wrote after the world crisis of the 70s:

Contemporary economics employs, for soothsaying, the most impressive black magic of all: science, out of the bowels of computer.... It will yield quantitative outputs when you feed it quantitative inputs.

But can the periodicity of world cycles embracing all the main groups of countries and industries of the capitalist economy, with the boundless specific character of the trends of development of each of them, in general be forecast on a more or less long-term basis? In contrast to capitalist political economy Marxist political economy gives an affirmative answer to that question. An infinite number of 'vectors of force' operate during each national cycle, and even more of each world one, and not only vectors of an economic order. A considerable number of them are not amenable at all to precise accounting and theoretically substantiated prediction.

Just the same Karl Marx managed in his day to create a scientific methodology for approaching long-term forecasting of the cyclic development of capitalism as a socio–economic system linked into a single economic organism by the world market. In the attempt at laying bare the laws by which crises of the market of the world are governed, not only their periodical character, but the exact dates of that periodicity must be accounted for. The distinctive features, moreover, peculiar to every new commercial crisis, must not be allowed to overshadow the aspects common to all of them.

In the course of these crises, moreover, in which, as Marx noted, the contradictions of all the elements of the capitalist process of production were reflected, the consequences could, in turn, become causes, and phenomena not originally regular, but apparently haphazard, later begin to have an increasing influence on the periodicity of the succession of the world cycle's phases.

Developing these propositions in subsequent work on the French edition of the first volume of *Capital*, Marx stressed:

So far the periodic length of these cycles is ten or eleven years, but there is no reason to think that figure constant. On the contrary, one may infer from the laws of capitalist production, such as we have just developed them, that it is variable and that the period of the cycles will gradually shorten.

These deductions, confirmed by incontestable historical facts, provide the initial methodological basis for analysing not only the past and current lines of development of the world capitalist cycle, but also of its prospective ones. As the productive forces have grown and been internationalised their time scale has in fact acquired a tendency to shrink in which is reflected, in particular, the deepening of the main contradiction of this mode of production.

Long-term analysis of the postwar cyclic fluctuations of production and international trade allows us to consider the onset of the world crises in the 70s to have been a completely natural phenomenon. In a situation of mounting inflation, acute aggravation of the monetary and financial, energy, raw material, and other world economic problems of capitalism, it was also not fortuitous that they affected the overwhelming majority of industrially developed countries and many developing ones. These crises, in which all the main contradictions of modern capitalism were manifested, in fact completed the ordinary world business cycles of the 70s, and laid the material basis for the next effect of the cyclic development of production in world business.

It is impossible, of course, to determine in advance, with the necessary reliability, just how this process will develop in the 80s, let alone the 90s, whether the regular cyclic crises will^become more destructive or shallower compared with the corresponding phases of preceding cycles, and whether it will be expressed in an absolute decline in the main annual indices for the whole world economy, or in only a sharp relative fall of them. All that will depend on the concrete combination of the most varied factors, including whether the various countries and spheres of the world economy enter the crisis simultaneously or separately, at what time in one calendar year or another the boom phase will be completed and the fall in production growth rates begin in the main cycle-forming industries. That will also depend on the further effect on their course of development of such crisis processes of a non-cyclic character as the primary commodity, energy, and foreign exchange crises, unchecked inflation, etc. The influence and dynamics of the growth of socio-political contradictions, and of the most important trends in the evolution of the international situation, will also affect the shaping of the features of the next world cycles.

But, allowing for all these factors, one thing is certain: 220 new cyclic explosions of the world capitalist economy are quite to be expected and inevitable. Modern capitalism, though employing the whole arsenal of state-monopoly measures, is unable to avoid them.

The postwar development trends reviewed above allow us to suggest that the world economic cycle of the second half of the 70s will in all probability be completed at the turn of the decade by a new crisis phase, which will probably mark capitalism's entry into the next cycle, which can be expected to last, judging by the trends considered above, until the middle or the early second half of the 80s.

The threat of even more powerful economic upheavals than the 1974–75 crisis will continue in the future. In any case the now rapidly increasing instability of the capitalist economy, and especially of the system of world economic relations, in the present stage of the general crisis of capitalism is more and more extending the objective premises for the probability of such upheavals with serious repercussions on the living standards of peoples and therefore fraught with serious social consequences for the capitalist social system.

Study of the patterns of postwar cyclic movement in no way excludes, of course, the need for a systematic investigation of the concrete phenomena and inner logic of this movement in individual countries and industries in the context of the world cycle, but on the contrary presupposes it. But a truly all-embracing idea of the beginning of the next world cycle and of the timing of the succession of one phase by another, can mainly be got by a combined, interlocked analysis both of the long-term economic processes in capitalist countries and of the system of world economic relations as a whole in the given concrete situation.

This kind of analysis is particularly urgent today when the internationalisation of production and level of the international division of labour have reached unprecedented heights. The patterns of the dynamics and periodicity of the cyclic fluctuations of any country's economy taken separately are now almost impossible to appreciate and comprehend simply from studying its internal trends of development, leaving aside their interaction with similar tendencies in other countries, and so with the long-term patterns and current peculiarities of the movement of the capitalist cycle in general. It has hardly been justified, moreover, to limit this interaction in recent years simply to the context of the industrial centres, leaving out of account the peculiarities of the effect of the break-up of the colonial system on the world cycle and the specific features of the postwar cyclic development of the periphery stemming from that.

The statistics cited above indicate that the acceleration of developing countries' industrial growth rates has so far, as a rule, encouraged a certain smoothing out of the depth of world recessions of production. During the whole cyclic development of capitalism since World War II their annual rate of industrial production, and of the GDP as a whole, has not once fallen in absolute terms below the level of the preceding years, although the range of fluctuations of their growth rates between the high and the low of each world cycle has been very considerable. The expanding markets of the developing countries at the same time favoured a shortening of the passage from the crisis phase to the boom phase.

This influence of the consequences of the break-up of the colonial system on the course of the contemporary world cycle may probably be maintained to some extent in the immediate future as well, but it is most likely to prove passing. Given further intensification of the immanent laws of the capitalist mode of production in developing countries, their economic development may already be transformed in the distant future from a smoothing factor to one consistently aggravating world capitalism's cyclic fluctuations.

The development trends of the fundamentally new industries whose origin is directly linked with the advances of the scientific and technical revolution, and in no small degree with the unprecedented increase on that basis of militarisation of the economies of many capitalist countries, will in all probability operate in the same direction. The new chemical and manufacturing industries (electronics, aerospace, most branches of military production largely based on budget appropriations and government orders, etc.) played a damping role in one way or another in the first postwar decades, and still do so, in the cyclic production recessions of the industrial centres, but in recent years this role has not been so strong, and has naturally begun to decline.

The development of the technically more advanced industries, also subject to the operation of the objective laws of the functioning of capitalist production, ultimately leads to further growth of its inner contradictions that govern the dynamics and scale of world business cycles. The facts of the absolute decline in the level of production during the crisis of the mid-70s in the technically more perfected and newer industries that had continued to show quite high growth rates during the preceding postwar overproduction crises points to that. There are solid grounds for suggesting a heightened role of industries of this kind in the cyclic phases of the economy of modern capitalism.

When we make a comprehensive study of all these trends so important for understanding the structural shifts being considered and the crisis processes in the world capitalist economy, it would be wrong, however, to ignore the gradually appearing prospect of limiting the sphere of operation of trends of this kind on the world capitalist cycle within the context of the world economy because of the probable extension in coming decades of the objective premises for consolidating a mutually advantageous and stable division of labour between socialist countries and the countries of the nonsocialist world. The acuteness of the cyclic production crises may apparently be attenuated to some extent for those capitalist and developing countries that take the road of strengthening comprehensive economic and technical cooperation with countries with planned economies in every way possible. The patterns of the socialist mode of production and its international relations will inevitably have an increasing effect on world economic relations in the last quarter of the twentieth century.

As the effect of socialism and the scientific and technical revolution on the growth and internationalisation of social production increases on a world scale, the task of studying the cardinal problems of the cyclic development of the capitalist system as a whole, and of its component parts, will undoubtedly become more complicated and broadened. And so long as this system exists these tasks will remain of paramount importance for study of its antagonistic contradictions, which are graphically manifested in world cyclic crises of overproduction.

Bibliography

Chand, Ramesh (2003) *"Impact of Trade Liberalization and Related Reforms on India's Agricultural Sector, Rural Food Security, Income and Poverty."* Mimeo. Paper prepared for the GDN meeting Delhi, 2004.

Chand, Ramesh: *Agricultural Diversification in India: Potentials and Prospects in Developed Region*, New Delhi: Mittal Publications, 1999.

Chrispeels, Maarten : *Plants, Genes and Crop Biotechnology*, Sudbury MA, Jones and Barlett Publishers, 2003.

David Sadava: *Plants, Genes and Crop Biotechnology*, Sudbury MA, Jones and Barlett Publishers, 2003.

Goodman, D.C.: *From Farming to Biotechnology: A Theory of Agro-industrial Development*, Oxford, Blackwell, 1987.

Jha, N.K.: *Bank Finance and Green Revolution in India*, Amar Prakashan, Delhi, 1985.

Metting, Jr., F.B.: *Soil Microbial Ecology: Applications in Agricultural and Environmental Management*, Marcel Dekker, New York, 1992.

Rao, M.G. and R.J. Chelliah: *Survey of Research in Indian Fiscal Federalism*, National Institute of Public Finance and Policy, New Delhi, 1991.

Shetty, Kalidas: *Food Biotechnology*, New York, Dekker/CRC Press, 2005.

Index

□□□